POPPER AND THE HUMAN SCIENCES

NIJHOFF INTERNATIONAL PHILOSOPHY SERIES

VOLUME 19

For a list of other volumes in this series see final page of the volume.

Gregory Currie and
Alan Musgrave, editors

Popper and the Human Sciences

1985 **MARTINUS NIJHOFF PUBLISHERS**
a member of the KLUWER ACADEMIC PUBLISHERS GROUP
DORDRECHT / BOSTON / LANCASTER

This is a volume in the subseries 'Contributions to Philosophy' under the editorship of J.T.J. Srzednicki. An earlier volume was published on the philosophy of John Wisdom (Nijhoff International Philosophy Series volume 17)

Distributors

for the United States and Canada: Kluwer Academic Publishers, 190 Old Derby Street, Hingham, MA 02043, USA
for the UK and Ireland: Kluwer Academic Publishers, MTP Press Limited, Falcon House, Queen Square, Lancaster LA1 1RN, UK
for all other countries: Kluwer Academic Publishers Group, Distribution Center, P.O. Box 322, 3300 AH Dordrecht, The Netherlands

Library of Congress Cataloging in Publication Data

Main entry under title:

Popper and the human sciences.

(Nijhoff international philosophy series ; 19)
Bibliography: p.
Includes index.
1. Popper, Karl Raimund, Sir, 1902- --Addresses, essays, lectures. I. Currie, Gregory. II. Musgrave, Alan. III. Series: Nijhoff international philosophy series ; v. 19.
B1649.P64P67 1985 192 84-1165

ISBN 90-247-2998-X (Hardbound)
ISBN 90-247-3141-0 (Paperback)
ISBN 90-247-2331-0 (Series)

PRINTED IN THE NETHERLANDS

CONTENTS

PREFACE

Although Sir Karl Popper's contributions to a number of diverse areas of philosophy are widely appreciated, serious criticism of his work has tended to focus on his philosophy of the natural sciences. This volume contains twelve critical essays on Popper's contribution to what we have called the 'human sciences', a category broad enough to include not only Popper's views on the methods of the social sciences but also his views on the relation of mind and body, Freud's psychology, and the status of cultural objects.

Most of our contributors are philosophers whose own work stands outside the Popperian framework. We hope that this has resulted in a volume whose essays confront not merely the details of Popper's arguments but also the very presuppositions of his thinking. With one exception, the essays appear here for the first time. The exception is L.J. Cohen's paper, which is a revised and considerably expanded version of a paper first published in the *British Journal for the Philosophy of Science* for June 1980.

We would like to thank Loraine Hawkins and Jane Hogg for their editorial assistance and June O'Donnell for typing various manuscripts and all the correspondence which a volume of essays entails.

University of Otago
Dunedin, New Zealand

G. Currie
A. Musgrave

1. THIRD WORLD EPISTEMOLOGY

L. Jonathan Cohen

1. Sir Karl Popper's striking hypothesis about a third world of objective knowledge deserves careful scrutiny because it is certainly his boldest conjecture. Moreover the hypothesis asserts an ontological thesis which can easily be adapted to fit certain non-Popperian methodologies, such as inductivism. So the hard core of the hypothesis is not hit by standard criticisms of Popperian methodology, and needs separate consideration. If it survives criticism, it will stand as a rather powerful theory about the nature of human culture: if it does not, then perhaps there is something to be learned from the reasons for its failure.

2. In Popper's view the world of objective knowledge, as he calls it, or more generally the world of 'the objective spirit', is man-made[1]. But it exists to a large extent autonomously, because the moment our critical and creative thinking has produced a theory that theory may itself produce further problems. It follows that when any of these further problems is formulated by a human being, as in Goldbach's conjecture, the problem has been discovered, not made, by us[2]. Indeed a large part of the objective third world of actual and potential theories and books and arguments arises as an unintended by-product of the actually produced books and arguments and of human language[3]. There is thus a constant feed-back between ourselves and our work. It follows that the main aim of all historical understanding is the hypothetical reconstruction of a third-world problem situation. For example, how is Galileo's theory of the tides to be understood? In order to achieve such understanding, says Popper, we have to investigate a fairly complex third-world object[4]. The object consists not just of the standard problem about the cause of tidal motion, but of this together with its then

Currie, G. and Musgrave, A. (eds.), Popper and the Human Sciences.
 ISBN 90-247-2998-X

current background of argument about the truth or falsity of Copernican theory, within a certain given theoretical framework of conservation laws.

It follows that the contents of the third world are rather diverse. Some, according to Popper, are just what Frege meant when he talked about the objective contents of human thoughts[5]. Indeed Popper tells us that 'theories, or propositions, or statements are the most important third-world linguistic entities'[6] But we are told also that the third world contains poetic thoughts as well as scientific ones[7]. Indeed it contains works of art, theoretical systems, problem situations, critical arguments, discussion-states, the contents of books and journals, and also scientific conjectures or hypotheses, experiments and their evaluations[8]. Moreover, just as the elements of the third world are thus richly diverse in character, so too are the processes in which they participate. There is not only a process of growth whereby a first problem gives way via tentative theorizing and error elimination to a second problem[9]. There is also competition and conflict between theories and problems, aspects of conjectures, interpretations and philosophical positions, and there are also comparisons, contrasts, analogies, problem-shifts, and so on[10]. Nor are all the objects in the third world of equal merit, since some objective knowledge is true, some false, some useful and some useless[11]. Indeed, while many theories in the third world are consistent, some – like Frege's *Grundgesetze* – are inconsistent, according to Popper[12]. As Popper puts it[13]:

> All work in science is work directed towards the growth of objective knowledge. We are workers who are adding to the growth of objective knowledge as masons work on a cathedral.
> Our work is fallible, like all human work. We constantly make mistakes, and there are objective standards of which we may fall short – standards of truth, content, validity, and others.

But the actual pattern of knowledge-growth is a Darwinian, evolutionary one. The fittest hypothesis survives[14].

Nor should the reader suppose that Popper is just indulging in an enormous metaphor or simile. He makes it quite clear by the comparison of his hypothesis with Plato's theory of forms, with Bolzano's theory of propositions in themselves, and with Frege's theory of objective thoughts, that he is ontologizing[15]. Indeed, he says in so many words that he is philosophically a 'realist' not only in regard to the physical world and the world of states of consciousness, but also in regard to the third world of objective knowledge[16]. His theory is meant to describe how knowledge really grows: it is not meant metaphorically[17]. Unfortunately, however, despite the length at which Popper has written about his third-world hypothesis, a number of rather important issues remain unclarified.

3. First, there is a problem about inter-communication. If intelligent creatures exist elsewhere in the universe with whom we terrestrial humans will never be able to communicate, the progress of objective knowledge may have gone much further than we shall ever know. If so, no human investigator will ever be able to test whether he has correctly conjectured the objective problem-situation in a particular area of enquiry at a particular time. Indeed, exactly the same problem arises in regard to any two cultures in the terrestrial past, if distance, xenophobia, geographical ignorance, or navigational incompetence prevented them from communicating information to one another. It follows that you cannot explain Galileo's theory of the tides, for example, by reference to the objective problem-situation in Galileo's time but only by reference to Galileo's subjective belief about that problem-situation. The explanatory goal of the third-world hypothesis seems unattainable.

One obvious solution of this difficulty would be for Popper to postulate a different world of objective knowledge for every closed group of intercommunicating intelligences. This would certainly deal with the problem across space, as it were. But, in view of the possibility that any such group of intercommunicating intelligences might be the outcome of past mergers or might give rise to future splits, Popper would also have to provide for corresponding processes of merging and splitting in his plurality of worlds of objective knowledge. And if the number of these worlds could thus fluctuate the conception of a third world of objective knowledge would hardly be on all fours with the conception of a physical world or of a world of states of consciousness. It is not now the conception of a totality of entities of a certain ontological category, but comparable rather to such less philosophical conceptions as that of 'the world of ancient China' or 'the world of the remoter galaxies'.

Accordingly, if the ontological unity of the third world is to be maintained, it would be necessary instead to suppose a potentially divided world, wherein perhaps many distinct cathedrals of objective knowledge were in process of mutually independent erection and were at different stages of growth. But the autonomy of the third world is now severely limited. Its ability to generate creative problems is now restricted by uncrossable frontiers between some of the objects that it contains. And these uncrossable frontiers are not the outcome or the explanation of any intrinsic epistemological factors. They are merely postulated *ad hoc* as the counterparts of certain physical barriers between groups of intelligent creatures.

I do not know which of these two possible responses would seem least objectionable to Popper. But his third-world hypothesis can hardly come to grips with the problem of intercommunication without losing at least some of its initial simplicity and metaphysical glamour.

4. There is a worse difficulty, however. 'Every scientific theory', says Popper, 'has, objectively, a huge set of important consequences, whether or not these have as yet been discovered'[18]. That is how the problem about Goldbach's conjecture came to exist in the third world long before it was actually discovered. And the third world also contains inconsistent theories, like Frege's *Grundgesetze*. But according to classical logic a contradiction implies anything whatever. So whatever proposition you care to mention is already derivable within some third-world theory. It looks as though the masons who are building the grand cathedral of objective knowledge have not only completed it but also rather disastrously over-fulfilled their norms.

How can this difficulty in Popper's hypothesis be circumvented? It is tempting to suppose that, in relation to every theory contained by it, the third world also contains the fact that the theory is consistent or inconsistent, as the case may be. Indeed, Popper himself writes of 'the objective hitherto unnoticed fact, belonging to World 3, that there was an inconsistency in Frege's theory ....'[19] After all, if the third world is attributed the capacity, as it were, to generate logical consequences, the capacity to detect inconsistencies might just as well be attributed to it. Then in accordance with such detections inconsistent theories can be consigned to some special category of demerit within the third world. But has the third world a solution for the decision-problem in regard to polyadic quantification-theory? No, for presumably Church's theorem is to be interpreted as a proof that the third world cannot contain a solution to this problem. It follows that the third world can hardly be supposed uniformly capable of including, whenever a new theory is fed into it, the fact that it is consistent or inconsistent, as the case may be; or we should be attributing to the third world a capacity that is demonstrably impossible.

It may seem preferable therefore to suppose that the third world operates with a relevance logic or with some other criterion of logical consequence that filters out any logical consequences that are due purely to the inconsistency of their premises. But any such supposition seems to run counter to the basic conception of the third world as including *all* theories, since Gentzen's theory of natural deduction, for example, is a classical logic and its presence in the third world would seem to enable inconsistencies there to entail anything whatever. Nor, presumably, would Popper want to suppose that each theory in the third world is insulated somehow from every other, so that Gentzen's logic could never be used to deduce paradoxical consequences from Frege's *Grundgesetze*. Such a supposition would be tantamount to supposing that there are as many distinct third worlds as there are distinct theories. And it would also destroy the inner unity of the third world which is required to account for its autonomous creativity.

Nor would it help at all if Popper were to exclude internally inconsistent theories altogether from his third world. The trouble is that in order to remain sufficiently distinct from Popper's second world (the world of mind) the third world has to contain propositions and their interrelations, rather than assertions of these. Hence within the third world no distinction may be drawn between *my* assertion of a certain proposition or yours or Frege's. Though propositions or theories get there as a result of our intellectual activity in the first and second worlds, once they get into the third world they operate in their own right, as it were, and Popper insists that false propositions may be there as well as true ones. It follows that the third world may contain very many pairs of mutually inconsistent propositions supplied by different thinkers, and the co-presence of any such pair has presumably just as much right to be considered a conjunction (or a pair of premisses for deduction) as the co-presence of any other propositions that strike sparks off one another within the autonomous creativity of the third world.

A plausible line of defence here might seem to be for Popper just to acknowledge that at any one time the third world might indeed contain a contradiction and consequently be over-full. He could then argue that from time to time the process of theory elimination would tend to cut it down to size. At any stage at which all inconsistencies had been eliminated the undesirable population explosion would be damped down. But this solution of the problem would leave us with the paradox that, each time an inconsistency returns to the third world as a result of human intellectual productivity, the third world reverts to an identical state of excess in regard to its population of propositions. Indeed, the third world can apparently be made to revert to that state even by sheer mischievousness. All you have to do is to contradict someone on some debatable issue, and thereby you sabotage the propositional selectivity of the third world until the debate is resolved.

In sum, when one presses the idea out into the requisite fine print, it is very difficult to see how Popper can maintain the autonomous creativity of the third world without getting involved in unacceptable paradoxes about the consequences of inconsistency. Yet without this autonomous creativity Popper's conception of a third world could make no claim to possess explanatory potential. It would merely gloss the actual record of second world developments.

5. A further difficulty arises as a result of Popper's claim that the third world contains all the consequences of any scientific theory, whether or not these have as yet been discovered. For it seems that the third world may thus contain not only the problem for which Goldbach conjectured his solution but also the correct solution. If it is derivable from the postulates of number theory that every even number is the sum of two primes, or that some even number is not such a sum, then one or other of these two theorems ought already to be in the third world. It got there just as soon as the postulates did. Of course, the proof may not be there yet, and perhaps will never be. And perhaps, after all, Goldbach's theorem is neither provable nor disprovable. But clearly Popper's hypothesis entails that the third world contains an enormous amount of mathematical knowledge – in the form of provable but unproved theorems – which is as yet unknown to any human being. It follows inevitably that the growth of such knowledge cannot be a third-world process. In this respect at least the growth of human knowledge can only be represented by a growth of subjective awareness, as human beings come gradually to discover provabilities that are already (according to Popper) part of objective knowledge. Correspondingly, the problem-situation at any particular time has to be viewed in subjective terms, not in objective ones, since it is only in human minds, not in the third world, that any of the relevant problems are as yet unsolved.

Nor is this just a difficulty for the Popperian epistemology of mathematics: it affects the history of natural science also. The apparent discrepancy between Newton's theory and Flamsteed's observations, in regard to the motion of the lunar apse, created an important problem-situation in early eighteenth-century mechanics[20]. Even Clairaut at first thought that the motion of the lunar apse was wholly inexplicable by Newton's laws, and started to formulate a new hypothesis regarding gravitation. But eventually he was able to carry out calculations showing that the observational results were indeed predictable from Newton's laws. Now according to Popper's hypothesis, if Newton's laws, plus the relevant statements of initial conditions, were already in the third world when Clairaut began his investigations, then so too was the fact that the appropriate motion of the lunar apse is derivable from these premisses. So, if the problem-situation that Clairaut encountered had really been determined by the third-world situation, he would never have wasted any time on trying to develop a rival hypothesis to Newton's and would instead have concentrated on his new calculations from the start. The only way for a historian to understand what actually happened at first is by reference to Clairaut's subjective but mistaken belief about the problem-situation.

It would be open to Popper to exclude this kind of difficulty in this third-world hypothesis by revising his view so as to deny that the logical consequences of a theory enter the third world before some human scientist discerns them. But then the autonomy of the third world would be gravely diminished and thereby, as remarked earlier, its claim to possess explanatory potential would be undermined. The inconsistency in Frege's *Grundgesetze*, for example, would not have entered the third world until Russell discovered it. So it wouldn't have been the objective autonomy of the third world that threw up the inconsistency, as it were, but the subjective creativeness of Frege's mind. Equally, Goldbach would have to be given the credit (which he surely deserves) for having posed the question that his conjecture was designed to answer. All that would have pre-existed Goldbach's question was the terminology in which it was formulated, along with some other (presumably suggestive) questions about similar topics.

6. It might be urged, on behalf of some supposed nugget of truth in Popper's hypothesis, that even when one speaks of, say, Clairaut's mistaken belief about the problem-situation, one assumes the existence of an objective problem-situation for Clairaut to misconceive. But if the third world is to contain objective problem-situations, as distinct from misconceptions of them, it must presumably contain at any one time the currently best possible evaluations of whatever theories are

under consideration at the time. The comparisons and appraisals that enter into a particular problem-situation must be those that ideally rational scientists would make, on the basis of available evidence, rather than those that fallible humans like Clairaut would make. Presumably no detectable errors would enter the third world. But then Frege's *Grundgesetze*, for example, would have been barred at the outset because of its inconsistency, contrary to Popper's express stipulation, and there would be no explanation, within the third world, for Russell's discovery. A history of the third world would be like that kind of misleading history of science which skips blithely from one peak of success to another and overlooks most of the errors and half-truths in the valleys below. (Some related difficulties about the connections between Popper's three worlds are discussed by Currie [1978].)

7. Finally there is serious difficulty about Popper's claim that his epistemology is 'a largely Darwinian theory of the growth of knowledge'[21]. On this view theories compete for survival and the fittest survives.

But which are the species and which are the species-members in this epistemological Darwinianism? If theories are cast in the role of species, then presumably the species-members are the representations of these theories in the minds and writings of individual scientists. But if that is so then the real struggle for survival takes place in the first and second worlds, not in the third. And yet a refuted theory is not automatically deleted from all the books in which it was printed. Many tokens of it remain, even when its more successful rivals enjoy much greater popularity. It follows that competition between theories cannot be envisaged as a Darwinian struggle in which theories operate as species and theory-representations as species-members. Are, then, the theories themselves to be conceived of as species-members, competing for individual survival in the third world? But if so, they seem for the most part rather ill-suited to constitute species. They serve their intended purposes only in virtue of being distinctively tailor-made for these and therefore different from one another. In sum, one is inevitably driven to the conclusion that any talk about Darwinain evolution in Popper's third world is either false or just metaphorical, contrary to Popper's own claim[22]. The complaint here is not that developments in Popper's third world are not the result of genetic mutations or physiological adaptations. How could they be? The trouble is rather that the central feature of the evolutionary process has no counterpart in the third world. Speciation does not play any important role in third-world development. Yet it was what Darwin wrote his great book about.

8. Actually, the situation is even worse than this. The proposed analogy between Darwinian evolution and the progress of knowledge does not break down merely over the question of speciation. It suffers from several other flaws. Nor should this be at all surprising. Darwin's theory was originally put forward as an alternative to those theories which sought to explain adaptedness by reference to some kind of rational design. Instead of appealing, like Christian theology, to God's rational plan, it contrived a pattern of explanation, via natural selection among random mutations, which had no need to assume the occurrence of any rational act. And not only was Divine rationality – i.e. the application of Divine knowledge for the fulfilment of Divine purposes – irrelevant to Darwinian explanation: no other kind of rationality was allowed to have a part in it either. In particular, no heritable innovations were to be traced back in Lamarckian style to learned adaptations. So no evolutionary change of any kind came about through the application of intelligence and knowledge to the solution of a problem. But by these standards science, in striking contrast, is a veritable paradigm of rationality. Scientists use their intelligence and previously acquired knowledge to discover more about how things happen, and they often learn thereby also how to prevent things from happening. Every stage in the progress of science is the product of intellectual effort and builds consciously on previous intellectual achievement. It seems quite inappropriate therefore, if not deliberately perverse, to think of this progress as a species of Darwinian evolution – as if something that is essentially rational could be a species of something that is essentially non-rational.

Of course, if you ignore the conception of rationality that contrasts with natural selection and suppose that rationality operates only via mechanical heuristics, then even science, or most of it, will seem non-rational. There is certainly no mechanical procedure for discovering explanatory hypotheses in unfamiliar fields of investigation. The main rationality of science consists, however, not in the partly unconscious processes that result in the production of a hypothesis or family of hypotheses but rather in the conscious, intelligent and well-informed application of appropriate criteria for selecting problems, appraising experimental designs, and evaluating hypotheses. Correspondingly, natural selection is a non-rational process because it involves neither consciousness nor intelligence nor the application of knowledge. That was at the heart of Darwin's idea. That was what shook the nineteenth century theologians. That is what sets Darwinian evolution utterly and irremediably apart from the rational processes involved in scientific enquiry. So any attempt to construe science as an evolutionary process can be expected to founder in a sea of incongruities.

We have already seen that one of these incongruities is the fact that science contains no obvious analogue for evolutionary speciation in nature. In terms of Peirce's type-token distinction, it is theory-types that get attacked by counter-evidence, not theory-tokens. Yet in nature it is the individual members of a species that get wiped out by ecological hazards. So, in principle, one refutation of a hypothesis may be sufficient, so long as it is replicable, to demote the hypothesis within Popper's third world. But many individual members of a species need to be prevented from reproducing themselves if the species is to be wiped out. Again, it is perhaps the ideas and principles constituting a theory that are supposed to correspond to the phenotypes instantiated by a biological population. But what corresponds to the gene-pool then? Or if the ideas and principles are supposed to correspond to the gene-pool, and some general characteristics of theories are supposed to correspond to biological phenotypes, then why is it that the marriage of two mutually consistent theories produces a quite determinate offspring? That is, exactly the opposite happens to what the comparison with genes suggests: the set-up is deterministic, not probabilistic.

A further pair of incongruities arise in relation to the proposal of hypotheses.

First, the nature of the hypotheses proposed is strongly affected by factors of the kind that also control preferences between hypotheses. A scientist who has read up the literature in his chosen field of research has a mind that is well-stocked with factors that will affect the content of the hypotheses that occur to him, and at least some of these factors – in particular, beliefs about previous experimental results – will also be relevant to his evaluation of those hypotheses. The evolutionary analogue would be for the nature of genetic mutations to be causally affected by the kind of environmental pressures that are operating. But any such coupling is excluded by evolutionary theory. It may have a place in Lamarckian theory, but not in Darwinian. Similarly, cross-lineage borrowing is not an evolutionary mechanism, but scientists in one field often exploit experimental techniques or mathematical procedures that have been developed in another field.

Secondly, if scientists actually tried to take survival value into account when they put forward their hypotheses, they would aim at hypotheses which were very difficult to falsify. Cowardice, not boldness, would bring rewards. If instead they followed Popper's advice and aimed at hypotheses which were in principle highly falsifiable, they would be doing exactly the opposite of what an evolutionary interpretation of science requires. In particular, if their hypotheses should normally generate novel kinds of testable prediction, as Popper – like Bacon – recommends, this would have to be interpreted in terms of the evolutionary analogy as an unnecessary courting of danger from environmental hazards. Popper's falsificationism is deeply at variance with his evolutionism. Perhaps it will be said that evolutionism here is to be construed simply as an account of what does happen in the progress of science, not of what ought to happen: it does not make normative recommendations to scientists and therefore what it says cannot be in conflict with any recommendations. But even if we admit this it will not eliminate the incoherence from Popper's philosophy, since his falsificationism is not just a normative recommendation. Falsificationism is supposed also to represent what the mainstream of science has always been like.

Finally, one might well ask, in regard to Popper's evolutionary interpretation: what new research into the history or sociology of science has it stimulated or could it stimulate? The answer seems to be: none. The actual similarities of structure between scientific progress and biological evolution are so few and so slight, and are counterbalanced by so many important dissimilarities, that the evolutionary interpretation is obviously much more likely to befog the historian or sociologist than to shine a search-light ahead of him.

Of course, none of this in any way counts against the conception of human science as a factor in the *biological* struggle for survival. It is highly plausible that some kinds of deductive and inductive mechanisms, and some innate quality-space, are a part of humanity's genetic heritage. It is also at least plausible to hold that modern science constitutes the most effective method of putting that innate endowment to work in the pursuit of knowledge. But nothing follows from this that would allow us to treat the actual progress of science as an evolutionary process. No *natural* selection operates, so far as we can tell, to bring about changes in scientific procedure. The time-scale is grossly inappropriate. Science can make vast progress within a single year of a man's life, whereas any evolutionary changes in human biology would require several generations to become manifest. Nor is it even clear yet that, in the long run, science actually has got survival value. Perhaps a species' possession of a certain kind of innate cognitive endowment inevitably leads eventually to the development of nuclear physics, and this – given normal animal aggressive instincts – eventually leads a species to self-destruction. *Ipso facto*, however, the human species may well be unable to survive long enough for any of its members to know for certain either that this is the case, or that it is not the case. So even at this quite general level nothing definite seems to be gained by thinking of science within the conceptual framework of Darwinian theory.

9. We are therefore driven to reject Popper's hypothesis about a third world of objective mind in Darwinian evolution. Critical analysis reveals far too many paradoxes and incongruities for it to afford an acceptable theory of cultural development. Somehow this development has to be accommodated within a more restricted (monist or dualist) ontology[23].

Notes

1. Popper [1972] p. 147.
2. *Ibid.* pp. 160 – 161.
3. *Ibid.* p. 117.
4. *Ibid.* pp. 170 – 175.
5. *Ibid.* p. 109.
6. *Ibid.* p. 157.
7. *Ibid.* p. 106.
8. *Ibid.* pp. 106 – 111.
9. *Ibid.* p. 121.
10. *Ibid.* p. 165.
11. *Ibid.* p. 115.
12. *Ibid.* p. 57.
13. *Ibid.* p. 121.
14. *Ibid.* pp. 261 – 264.
15. *Ibid.* p. 106.
16. *Ibid.* p. 107.
17. *Ibid.* p. 261.
18. Popper and Eccles [1977] p. 140.
19. *Ibid.* p. 57.
20. Cf. Cajori [1962] pp. 648 – 650.
21. Popper [1972] p. 261.
22. *Ibid.* p. 261.
23. This is a revised and expanded version of a paper which appeared originally in *The British Journal for the Philosophy of Science*, 31 [1980] pp. 175 – 180.

2. PSYCHOANALYSIS, PSEUDO-SCIENCE AND TESTABILITY

Frank Cioffi

Karl Popper says psychoanalysis is a pseudo-science; others deny this. What is Popper's conception of pseudo-science and why does he hold Freud's psychoanalysis to be an example of one? What would it take to show that Freud's theory is pseudo-scientific?

Popper has characterized his demarcation theses as 'bold simplifications', and it is in this spirit that I will treat them and advance my own counters to them. Here is my 'bold simplification' of the substantive issue: when the *bona fide* empirical status of psychoanalysis is rejected, that is, when it is declared a pseudo-science, what is under discussion is not a theory but a practice. (Since a practice implies practitioners a certain indeterminacy necessarily characterizes such a claim). My 'bold' *exegetical* simplification is that this is Popper's view as well.

I will argue that though Popper is correct to say that psychoanalysis is pseudo-scientific, and correct to say that it is unfalsifiable, he is mistaken to suggest that it is pseudo-scientific *because* it is unfalsifiable. There are empirical theses which are neither falsifiable nor pseudo-scientific, and others which are both. When Freud advances unfalsifiable theses he is being proto-scientific. It is when he insists that he confirmed (not just instantiated) them that he is being pseudo-scientific.

There are three main ambiguties about falsifiability as a criterion of demarcation between science and pseudo-science.
(1) Whether the mere possession of falsifiers is a sufficient condition of scientific as opposed to pseudo-scientific status, or whether more is required. (2) Whether the lack of falsifiers is a sufficient condition of pseudo-scientific status, or whether such utterances are non-scientific, or proto-scientific rather than pseudo-scientific. (3) Whether we are being enjoined to capitulate to apparent discrepancies between the

Currie, G. and Musgrave, A. (eds.), Popper and the Human Sciences.
 ISBN 90-247-2998-X

theory's observational entailments and reality, or merely to search assiduously for such discrepancies and note them. Popper is sometimes ambiguous as to all of these but I shall argue that his considered view is, in each case, the latter[1].

Does Popper hold that a pseudo-science must be unfalsifiable, that is, is the lack of falsifiers a necessary condition of pseudo-scientific status? Adolf Grünbaum is foremost among those who have insisted on construing Popper's criterion for demarcating science from pseudo-science as formal falsifiability, that is, the mere possession of falsifiers. In 'Is Freudian Psychoanalytic Theory Pseudo-Scientific by Karl Popper's Criterion of Demarcation?' ([1979*b*] p. 131) Grünbaum says: 'according to [Popper's] criterion, the hallmark of the scientific status of a theory is that empirical findings which would refute it are logically possible.' In 'Is Falsifiability the Touchstone of Scientific Rationality?' (1976) Grünbaum repeats this characterization: 'According to Popper's falsification criterion, falsifiability is sufficient for scientific status.' (p. 223). 'For Popper the mere falsifiability of a hypothesis suffices for according scientific status to it.' (p. 228). It is clear from what Grünbaum says elsewhere, and from the requirements of his argument, that the term 'scientific' in these remarks is being used as antithetical to 'pseudo-scientific'. In the concluding paragraph of his 'Is Freudian Psychoanalytic Theory Pseudo-Scientific by Karl Popper's Criterion of Demarcation?' he writes: 'Cioffi (1970) has proposed to modify Popper's criterion along the following lines: A theory's refutability is quite insufficient for its scientificality. Even if its claims "are eminently refutable", Cioffi tells us, the theory's pseudo-scientificality can be assured by its advocates wilful employment "of methodological precedures that are calculated to lessen the uncovering of potentially falsifying evidence." ' I will attempt to show that the thesis Grünbaum attributes to me is not a modification of Popper's demarcation criterion but that it *is* Popper's demarcation criterion.

Here are the reasons for believing that Popper holds that a theory which is formally falsifiable may, nevertheless, be pseudo-scientific. On p. 81 of the *Logic of Scientific Discovery* [1968*a*] Popper lists four devices 'for attaining, for any chosen axiomatic system ... its correspondence with reality', the third of which he describes in the following terms: ' ... we may adopt a sceptical attitude as to the reliability of the experimenter whose observations which threaten our system, we may exclude from science on the ground that they are insufficiently supported, unscientific, or not objective, or even on the ground that the experimenter was a liar. (This is the sort of attitude which the physicist may sometimes quite rightly adopt towards alleged occult phenomena).' On the next page he refers to psychoanalysis. (Remarks with similar import occur in Popper's replies in the Schilpp volume. At one point he speaks of 'one vast class of immunizing strategies – those that blame the observation for not presenting the true state of affairs.' ([1974], p. 1005). At another he says, 'we can always deny the objectivity, or even the existence of refuting observations' (p. 983), along with an allusion to 'people who refuse to look through Galileo's telescope.')

On p. 82 of the *Logic of Scientific Discovery* he says: 'It is impossible to decide by analysing its logical form, whether a system of statements ... is a refutable system ... Only with reference to the methods applied to a theoretical system is it at all possible to ask whether we are dealing with ... an empirical theory.'

In the paper 'What is Dialectic' ([1969], p. 313) Popper says: ' ... it is most characteristic of the scientific method that scientists will spare no pains to test and criticize the theory in question.' This sentiment is incorporated in 'Remarks on the Problems of Demarcation and Rationality' (Lakatos and Musgrave, [1968], p. 94), where it is supplemented: 'What characterizes the scientific approach is a highly critical attitude towards our theories rather than a formal criterion of refutability; only in the light of such a critical attitude and the corresponding critical methodology approach do "refutable" theories retain their refutability.'

In the autobiographical section of the Schilpp volume Popper writes: 'What impressed me most was Einstein's own clear statement that he would regard his theory as untenable if it should fail in certain tests .... This, I felt, was the true scientific attitude, utterly different from the dogmatic attitude which constantly claimed to find "verification" for its favourite theories. The scientific attitude was the critical attitude which did not look for verification but for crucial tests ...' ([1974], p. 29).

Moreover, since the 'Oedipus effect' (a self-verifying prediction) is defined in terms of the prevention of conflict with falsifying observations (Popper [1969], p. 38), it can only be raised as an objection to theories which are formally capable of conflicting with such observations, that is, which are 'P-scientific', as Grünbaum tendentiously dubs them, so that 'P-scientific' can hardly be the antithesis of pseudo-scientific. (That it is the notion of spurious confirmations rather than of unfalsifiability with which Popper is working is indicated, too, by the epigraph from Trollope's *Phineas Finn* which introduces *Conjectures and Refutations*.)

I conclude that Grünbaum is quite mistaken to assign to Popper the view that formal falsifiability is sufficient to rebut a charge of pseudo-science. If I may venture a diagnosis of Grünbaum's error, it is that he fails to notice that Popper is working with two antitheses: one, between non-scientific (or proto-scientific, or metaphysical) assertions which are to be distinguished from scientific ones on the score of their falsifiability and another, between pseudo-scientific and non-pseudo-scientific assertions where the relevant consideration is the conditions under which claims to have confirmed the theory are issued.

Grünbaum's claim that the inductivist criterion of scientific status is more stringent than Popper's is based on the same misunderstanding. Popper and the inductivists are giving different answers to the question, 'under what circumstances is a positive instance of a universal generalization to be considered supportive of the truth of that generalization?' Grünbaum intermittently confuses this question with the quite distinct question, 'under what conditions is a putatively positive instance of a phenomenon to be considered a genuinely positive instance?'

Now it is quite clear that when Popper says that positive instances of white swans are not capable of supporting the universalization that all swans are white he is not speaking of spurious positive instances but of genuine ones. The difference does not lie where Grünbaum locates it but rather in this: if we consider a universalization which it has not been possible to subject to attempted falsification, but of which there are genuine positive instances and, *a fortiori*, no negative ones, then Popper's inductivist must assign some probability value to it while Popper would not. That is, Popper's question is: what measure of support should we permit genuinely positive instances to afford a universal hypothesis with respect to them under circumstances in which it is not possible to distinguish apparently negative instances from genuinely negative ones? Popper's answer to this question is, 'None'. And whether or not he is correct, it is with respect to *this* question that Grünbaum ought to have drawn the contrast between Popper's falsificationism and the inductivist position, and not with respect to the non-methodological (formal) issue, 'what makes an assertion a candidate for scientific status or assessment?', where the distinctiveness of

Popper's position lies rather in his refusal to accord such status to uncircumscribed singular existentials as contrasted with the inductivist's willingness to do so[2].

1. *Is unfalsifiability a sufficient condition of pseudo-scientific status?* Though it is important to make Popper's distinction between theories which are instantiable but not falsifiable and those which are both, it might be better not to make it by calling the non-falsifiable ones 'pseudo-scientific'. Nor does Popper always do so. Sometimes he refers to them as 'non-scientific', or 'metaphysical'.

> Psychoanalysis ... is an interesting psychological metaphysics (and no doubt there is some truth in it, as there is so often in metaphysical ideas), but it was never a science. There are lots of people who may be Freudian or Adlerian cases: Freud himself was clearly a Freudian case and Adler an Adlerian case. But what prevents their theories from being scientific in the sense here described is very simply that they do not exclude any physically possible human behaviour. Whatever anybody may do is, in principle, explicable in Freudian or Adlerian terms. Adler's break with Freud was more Adlerian than Freudian, but Freud never looked on it as a refutation of his theory ... *the theory was compatible with everything that could happen – even without any special immunization treatment*. (Schilpp [1974] p. 985)

Even in this predominantly formalist passage Popper slips in non-formalist considerations, for example, the remark that Freud would not consider the occurrence of Adlerian cases as falsifying his theory. But on a formalist interpretation of Popper's remark what choice would Freud have? He would not even be able to conceive of what an Adlerian counter-example might be. If 'Adler's break with Freud was more Adlerian than Freudian' then it cannot be the case that the theory was *a priori* 'compatible with everything that could happen.'

I construe this apparent discrepancy in Popper's characterization to imply that what is pseudo-scientific about psychoanalytic theory is not its unfalsifiability, *not* the theory's 'compatibility with everything that could happen' but the pretence (or the illusion) that this is not the case. And in the passage of his autobiography (Schilpp [1974], p. 29) where Einstein figures as the paradigmatic scientist because of his expressed willingness to abandon his theory if it conflicted with observations, the antithetical 'dogmatic attitude' is characterized *not* as that of entertaining unfalsifiable theories but as 'constantly claiming to find "verification" for its favourite theories', which is a very different thing.

There are further remarks of Popper's which imply that it is not mere formal unfalsifiability which is his ground for considering a theory *pseudo-* as opposed to merely *non-*scientific. When Popper is illustrating the behaviour typical of pseudo-scientists it is often not their dogged invocation of auxiliary hypotheses to evade falsification he is complaining of but something else: their treatment of putative falsifiers, not merely as non-disconfirmatory, but as positively confirmatory, on the strength of their ability to suggest ways of reconciling the apparent falsifier with the theory.

His complaint is not that they evade falsification but that they arbitrarily identify apparent counter-instances as positive ones[3]. For example, the famous Adler anecdote: ' ... as to Adler, I was much impressed by a personal experience. Once, in 1919, I reported to him a case which did not seem to be particularly Adlerian, but which he found no difficulty whatever in analysing in terms of his theory of inferiority feelings, although he had not seen the child. Slightly shocked, I asked him how he could be sure of all this. "Because of my thousandfold experience", he replied. Whereupon I could not help saying: "And with this new case, I suppose, your experience is now even thousand-and-one fold." ' Note that Popper is not merely objecting to Adler's facility in turning the force of a putatively negative instance but to treating his ability to do this as transforming it into a positive one: 'And with this new case, I suppose, your experience has become thousand-and-one-fold.' (Popper [1969] p. 35).

The distinctiveness of this objection emerges even more clearly in the next paragraph: 'What I had in mind was that his previous observations may not have been much sounder than this new one: that each had been interpreted in the light of "previous experience" and at the same time, counted as additional confirmation. What, I asked myself, did it confirm? No more than the possibility of interpreting a case in the light of the theory.'

Popper is not merely objecting to the possibility of interpreting a case in the light of the theory (he even observes elsewhere that this is true of any theory), nor even just to the refusal to capitulate to a putative falsifier, but to this 'counting as additional confirmation'. There is no necessity for the rejection of falsification to be accompanied by the treatment of the apparent counter-instance as further evidence in favour of the theory. Instances which are not blatantly positive could be treated as *sub-judice*, as Freud, in a rare moment of restraint, initially treated the relation of the war neuroses to his libido theory. Why should Adler have modified his theory on the basis of an apparently disconfirmatory report by Popper? But he could reasonably have been required to refrain from treating Popper's counterexample as confirmatory merely because he was able to account for it within the terms of his theory.

When Popper makes unfalsifiability the basis of a judgement that a theory is pseudo-scientific it is usually in the context of a discussion of support. For example, when he asserts the pseudo-scientific status of 'psychoanalysis itself' it follows the remark that 'real support can only be obtained from observations undertaken as tests (by "attempted refutations").' Popper points out that the availability of the notion of ambivalence to account for apparently discrepant behaviour would prevent a given instance from functioning as a falsifier. But the mere availability of the notion of ambivalence to account for apparent falsifications (and the same is true of contamination – 'the Oedipus effect' – and resistance as well) does not suffice to make the theory which contains it pseudo-scientific, for the theory may only contain the notion because the world contains the fact. However, where this is so no claim can be advanced that the theory has survived testing.

Having recognized the asymmetry between instantiation and falsification we can say that the scientist is someone who would not be content with this state of affairs rather than, as the formal criterion suggests, that he should take no interest in it *qua* scientist.

2. *Is psychoanalytic theory testable?* Grünbaum roundly contradicts Popper's claim that psychoanalytic theory is unfalsifiable. He does not pause to ask which theses Popper might have had in mind. Not even Popper's circumscription of his claim to 'psychoanalysis itself', an expression which Popper never clarifies, moves Grünbaum to recognize the need for elucidation before proceeding to the pleasure of contradiction. His argument might be paraphrased thus: 'There is a body of theses advanced by Freud. Popper has characterized these theses as unfalsifiable. But among these theses is one which states that in their childhood, hysterics were sodomized by adults. The statement that a child has been sodomized by an adult has observational implications; therefore Popper was mistaken to say that psychoanalytic hypotheses are unfalsifiable.' This is like maintaining the empirical status of Christology on the grounds of the falsifiability of 'suffered under Pontius Pilate.'

Is it too much to ask for a level of discrimination which recognizes that a body of theses might be conceptually miscellaneous and that some might be testable and others not? There are distinctively psychoanalytic theses which do answer to Popper's characterization of them as unfalsifiable[4]. The impression created by Grünbaum that the belief that there are untestable psychoanalytic theses is an idiosyncrasy of Popper's, due to his failure to undertake elementary exegesis, can be dissipated by considering instances of psychoanalytic discourse which have been characterized as untestable by non-Popperian commentators.

Consider an example given by William Alston: 'We find Freud saying things like:

> At the very beginning all the libido is accumulated in the id, while the ego is still in process of formation or far from robust. Part of this libido is sent out by the id into erotic object-cathexes, whereupon the ego, now growing stronger, attempts to obtain possession of the object-libido and to force itself upon the id as a love-object. (*The Ego and the Id*)

It is clear that this cannot be construed literally. The ego and the id are not different persons .... We can picture each as a cartoonist's model in which all personality functions but one have atrophied. This gives us a picture, but we still don't know what to do with it. We don't know how to test the statement, when to say that this attraction has taken place, what role it might have in the formation of neuroses, etc..... Until such jobs are done, we are merely amusing ourselves with fables.' ([1964] pp. 441 – 2).

There is another kind of psychoanalytic utterance which poses problems of testability. This time not because of its figurative character but because of the motivational complexity of the situations with which it deals. The problems posed by the notion of infantile oedipal impulses are quite different from those which arise in connection with the relation of the ego to the id. This is illustrated by the remarks of Sybille Escalona, a psychoanalytic researcher, on the reasons for the unfalsibiability of some of Freud's developmental theses:

> Suppose one wished to investigate the psycho-analytic idea that, at about the ages of three and four, little boys characteristically experience hostile and aggressive feelings towards their parents and more specifically their fathers, and that these hostile feelings are based on the wish to possess the mother and replace the father in the family constellation. Suppose also that you create an experimental situation which would activate the child's feeling towards his parents. For instance, the father might exert his authority by sending the child off to his own room when the child wants the remain with the parents. If, to

> substantiate the core of the oedipal hypothesis, the child were required to display frank anger against his father, or frank possessiveness towards his mother, things would be reasonably simple. However, no one would seriously believe that oedipal conflicts are at work only when the child obligingly acts out both his wish and his anger at the obstacle to wish-fulfilment. Instead, we assume that the child tends to defend himself against becoming aware of – or openly reflecting – aggression because it engenders anxiety. Thus, if the child gives daddy a good-night hug and insists that he, rather than mummy, tuck him in, this behaviour may also confirm our original hypothesis. His desire to have the father put him to bed rather than the mother could be the result of a fearful state, i.e. as long as the father is with him the little boy can be sure the father is not doing anything to harm him. On the other hand, or also simultaneously, it may be an act of aggression towards the father in that it separates him from the mother for the time being. Or yet again, it may be because the little boy fears that if mother puts him to bed her seductive powers will prove too much for him; he will then express his possessive love for the mother and try to take his father's place, and the omniscient father will punish him for it. The example could be spun out indefinitely, and it is safe to say that there is nothing our little experimental subject could possibly do, from withdrawal, to sudden intense interest in phantasy play, to asking for a cake, that cannot be regarded in the light of the assumption that he is reacting to an oedipal conflict situation. This being so, it is self-evident that nothing the little boy can do will confirm the original hypothesis, since the hypothesis would still be applicable if he had done the opposite instead .... Psychoanalytic theory is greatly in need of validation, yet it is the kind of theory incapable of validation by available methods ... and which require different methods of validation from those now known to us. (Escalona, [1952] pp. 11 – 41)

In his paper 'How Scientific is Psychoanalysis?' ([1977] p. 251) Grünbaum writes, 'despite attempts by some Freudians to immunize the developmental theory against P-falsification, logically that theory as such is falsifiable.' It would have been helpful if Grünbaum had given examples of the 'attempts by some Freudians to immunize the developmental theory against P-falsification' that he had in mind. I do not believe that there have been any such attempts not anticipated by Freud himself in expositions of the 'theory-as-such'. Freud's followers were no freer with the notion of the 'sexual constitution' than Freud himself was. And what is the sexual constitution but our old friend 'hereditary degeneracy' about whose explanatory pretensions Freud himself was so scathing at the outset of his career. And Freud's 'sexual constitution' is even more contemptible. It is tendentious as well as empty, for we are never told how contributions made to its developments by the *sexual* constitution of the child is to be distinguished from that of its *general* constitution. At this point in his expositions Freud invariably invokes a 'quantitative factor' whose operation is only to be discerned *post hoc*.

Grünbaum has the makings of a reply to these arguments for the unfalsifiability of certain psychoanalytic theses: his distinction between psychoanalytic theory 'in itself' and the response of its defenders to actual or hypothetical falsification reports.

Consider Grünbaum's contrast between the 'theory-in-itself' and the tenacity of its defenders: 'Popper's indictment of psychoanalysis as intrinsically not being P-falsifiable has unwarrantedly derived plausibility from his failure to allow for the following distinction: the (revocable) falsifiability of the theory-as-such in the context of its semantic anchorage is a logical property of the theory itself, whereas the tenacious unwillingness of the majority of its defenders to accept adverse evidence as refuting is an all too human property of those advocates. Thus lack of methodological honesty on the part of the defenders, and even on the part of its originator, does not necessarily render the theory itself unfalsifiable.' ([1979*b*] pp. 137 – 8).

It isn't clear why Grünbaum lays such emphasis on this point. It would only matter if, from the fact that the 'theory in itself' was falsifiable, it followed that it could not therefore be pseudo-scientific. But it doesn't. Otherwise we would not speak of astrology, or theories of racial superiority, as pseudo-scientific since they are manifestly falsifiable, and for many of us have been falsified, without this fact inhibiting us from continuing to describe them as pseudo-scientific.

On one natural construction of Grünbaum's expression, 'the theory in itself', even Popper's paradigm of an unfalsifiable thesis would not be such since 'all swans are white' is manifestly empirical, and is not made less so by the refusal of some zoologists to accept that it had been falsified by the existence of black swan-like creatures in Australia. We can get round this by regarding the classification of Australian black 'swans' as non-swans as an elucidation rather than an emendation of the original thesis, but this makes the expression, 'the theory in itself', much more problematic than Grünbaum shows any sign of recognizing. For if we say that 'the theory-in-itself' is to be taken as encompassing the original statement of the theory, plus restatements of it in the face of falsification, what shall we call the theory without such restatements? Grünbaum's criteria lead to inconsistent conclusions. The theory-in-itself is both falsifiable, since it was not divested of its natural implications, and unfalsifiable, since it was.

Freud often puzzles us, not by putting forward claims without falsifiers, but by putting forward other claims as well whose natural force is to cancel the falsifiers of the more straightforward ones. For example, only the sexual instincts are libidinal, but the self-perservative instincts are libidinal too; social feelings are disguises for sexual ones, but social feelings are de-sexualized derivatives of feelings that were once sexual and are not themselves sexual.

How temporally proximate to the main text must elucidations be to warrant the judgement that they belong to the theory-in-itself? And must they be individuated biographically so that only the author of the original theory may modify, and thus imperil, the scientific standing of the theory?

When Grünbaum writes, 'despite attempts by some Freudians to immunize the developmental theory against P-falsification, logically that theory as such is falsifiable' ([1977] p. 251), it is not clear what kind of immunization is intended. If it is the kind which Popper refers to as 'blaming the observations for not representing the true state of affairs', then Grünbaum's remark, though possibly true, is pointless in a discussion of the pseudo-scientific status of a theory as contrasted with a discussion of whether it has potential falsifiers. Moreover, much of the immunization of Freud's developmental theory is not a matter of merely rejecting reported falsifications, but one of emending the content of Freud's developmental claims so as to lower the testability of the theory.

Consider, for example, the following account by Brill, Freud's American translator, of what is meant by the term 'sexual' in Freud's claim that only *sexual* vicissitudes can bring about a neurosis: 'A child may suffer from a sexual disturbance if he is neglected by a parent, or maltreated by someone in authority over him. The loss of a parent may make the child nervous: it is a disturbance in the latter's love life.' ([1948*b*] p. 49). Nor was this extension of the term 'sexual' confined to infantile life. Elsewhere Brill wrote: 'Some time ago a young woman consulted me, because, as she said, she was extremely nervous; she declared that she suffered from insomnia, that her appetite was poor, and that she entertained peculiar thoughts. When asked what she meant by peculiar thoughts, she replied that she simply could not stomach her mother, who was constantly "getting in her way". Whenever, for instance, she wished to do something, however trifling, her mother stood in the way. There was nothing of love or sex as we commonly understand these terms, so that to one who is not accustomed to our viewpoint it would have shown nothing wrong sexually. But as there is a very vital and intimate relation between child and parent, there was really a disturbance in the love life of this young woman. So you see how different our conception of sex is from that ordinarily held.' ([1948*a*] pp. 28–9)

If Brill is correct in his account of Freud's notion of sexuality it is no wonder that, as Richard Wollheim puts it, 'he uncovered its utter pervasiveness, he demonstrated its utter ubiquity.' The only question is what, given this extension of the scope of the term 'sexual', there could have been left to 'demonstrate' and 'uncover'. Now, how is it to be determined whether Brill's enlarged conception of sexuality is part of Freud's theory, or whether we are to dismiss Brill's remarks as just another specimen of the 'epistemological excesses of Freud's disciples'?

Grünbaum's distinction between the theory-in-itself and the unwonted tenacity of its adherents, even where we can make it with confidence, is of very limited relevance. Consider Rex Mottram, a character in Evelyn Waugh's novel, *Brideshead Revisited*, who wants to marry a Catholic girl and is receiving instruction to that end. His Jesuit instructor, who suspects his good faith, relates the following exchange: 'I asked him: "Supposing the Pope saw a cloud and said, 'It's going to rain', would that be bound to happen?" "Oh yes, Father." "But supposing it didn't?" He thought for a moment and said, "I suppose it would be sort of raining spiritually, only we were too sinful to see it." ' It isn't clear which immunising tactic Rex Mottram was employing: whether he was saying, 'When the Pope says it will rain, it rains; the reports that it didn't are erroneous,' which is equivalent to Freud's argument from resistance, or 'When the Pope says it will rain, it rains in an extended sense of rainfall which captures the essence of the original in a way in which the idea that it didn't rain because it didn't vulgarly rain doesn't,' which is equivalent to Freud's extension of the notion of libido. But let us suppose Mottram had not qualified the claim that it rained with the word 'spiritually', and had unequivocally asserted that it rained and that reports that it hadn't rained were erroneous. We would then have a case where it would be appropriate to remonstrate, as Grünbaum does with Popper, with someone who inferred a formal deficiency in rainfall statements because they were open to the stratagem employed by Rex Mottram. If someone were to argue from Rex Mottram's defence of the Pope's infallibility (or rather his misconception of this) that there was something intrinsically defective about rainfall statements, then a reminder of the distinction between rainfall-statements-in-themselves and the behaviour of those who have asserted them would be pertinent. The situation with Freud's libido theory is very different. We do not have clear cut falsifiers coped with by blatantly *ad hoc* hypotheses but rather a highly indeterminate thesis with highly indeterminate accompanying, or belated, qualifications.

What can be meant by 'the theory-in-itself' where the theory in question is replete with jargon and neologism, for example, terms like libido, narcissistic libido, ego-libido and self-preservative libido? Freud's response to apparent falsifiers of his libido theory, like the influence of the Kaiser's withered arm on his martial propensities, or of the pathogenic role of injuries to self-esteem and of the dangers encountered in warfare, can't be distinguished from 'the theory-in-itself' for they tell us what is meant by expressions like 'narcissism' and how it is distinguished from 'egoism', i.e. how libido differs from interest.

Grünbaum gives as a further example of the falsifiability of psychoanalytic theory the view that phobias represent the displacement of repressed sexual fears. Grünbaum thinks this has been refuted by the success of behaviour therapists in curing phobias by de-sensitization techniques. Grünbaum is under the illusion that Freud did not acknowledge that neurotic symptoms could be successfully cured without resort to psychoanalytic therapy. Freud explicitly states in the 28th of the *Introductory Lectures* that hypnotic suggestion may succeed in effecting permanent cures ('at times everything fell out as one could wish; one obtained complete and lasting success with little difficulty.') So Grünbaum's falsifiers were not falsifiers for Freud. Nor for Ernest Jones. Jones said of the war neuroses, 'I see no reason whatever why a psychoanalysis should be undertaken in the majority of the cases for they can be cured in much shorter ways.' ([1923] p. 593)

How is it to be determined whether in the light of these facts we are to revise our conception of Freud's 'theory-in-itself', or whether we are to consider the 'theory-in-itself' falsified and treat Freud and Jones as guilty of unwonted tenacity? If the grounds being given for the judgement that a theory is pseudo-scientific is the behaviour of its advocates, then the issue is of a kind to make the logical status of the 'theory-in-itself' irrelevant. On the other hand, if the behaviour of its advocates is introduced into a discussion of the logical features of the theory, then it is likely that it is their elucidations and emendations of the theory when confronted by falsifiers which is being adduced, and this does bear on the empirical content of the theory and cannot be put aside by an allusion to the 'scientificality' of the 'theory-in-itself'.

3. *Must unfalsifiable theses be pseudo-scientific?* Escalona's account of the Oedipus complex shows that Grünbaum is mistaken to deny that Freud's developmental theses are untestable. But what follows from this untestability? Not, I would argue, that they are pseudo-scientific. A situation such as Escalona envisages regarding Oedipal feelings may actually prevail. Things may be as Freud says but difficult to detect. If we are sceptical about this, and wish to dissuade others from wasting resources searching for them, we ought to indicate this otherwise than by calling such speculation pseudo-scientific. It would deserve this epithet only if the speculation were accompanied (as it is in Freud and many Freudians) by the implication that it had been confirmed, (indeed confirmed so fully that discreditable psychological motives had to be invoked to explain why there were those who were still recalcitrant, for example, by Jacob Arlow: 'The existence of the relationships which give the characteristic quality to the Oedipal phase is not a matter of conjecture or reconstruction. It is a matter of observation. Why these observations were not made before Freud is another problem pertinent to the resistance to psychoanalytic findings.' (Arlow [1960] p. 209). Charles Brenner advances the same argument in his article on psychoanalysis in the *Encyclopaedia of the Social Sciences*.)

Since Popper concedes the genuineness of the pheonomenon of ambivalence (and presumably of resistance as well) it is unlikely that he is demanding that these concepts be expunged from the theory. What then is the psychoanalytic investigator, on Popper's view, to do? The answer is, I think, to refrain from issuing failed disconfirmation reports which, under the circumstances, would be a form of perjury. What the situation calls for is not the abandonment of the theory, for example, of the Oedipus complex, but the candid admission that we are not as yet in a position to distinguish genuine from spurious negative instances. If Freud had been consistently and candidly nonfalsificationist there would be no occasion to upset the literati by calling him a pseudo-scientist.

What is, I think, an appropriate response to this feature of some psychoanalytic theses is provided by Roland Dalbiez in his book on Freud: 'Sooner or later we are induced to wonder whether besides instances in which the existence of a sexual activity in the child may be diagnosed from clinical signs there are not other instances of ''unapparent'' sexual activity, which can only be reached by means of psychoanalytic interpretation. It seems to us more likely that this question must be answered in the affirmative. But then a new question arises. If besides instances of apparent sexual activity, there are instances of unapparent sexual activity, are there still further instances of children without any psychosexual activity at all, whether apparent or ''unapparent''? Here we must merely confess our complete inability to settle the question one way or another ... the absence of exact sexual manifestations in childhood seemed to us as normal as their presence ... but it is quite clear that improvement of the psychoanalytic methods of exploration may ... enable us to demonstrate the presence of a genuine sexual activity which clinical ''macroscopic'' investigation failed to reveal. We cannot foretell the future of science.' (Dalbiez [1941] p. 255). Dalbiez sees that the Freudian concept of infantile psychosexuality is unfalsifiable but does not conclude that it is pseudo-scientific on that account.

A less indulgent attitude than Dalbiez' and one closer to Popper's is that expressed by R.S. Woodworth seventy years ago: ' ... the doctrine of Freud suffers the disability that it apparently cannot be put to a crucial test; for which ever way the test came out, the Freudian would find in the result a confirmation of his views. For example, a dream is always the expression of a repressed wish; but if a particular dream that is brought forward seems not to be the expression of a wish, it can be regarded as expressing the wish that the Freudian doctrine be not confirmed, or as expressing a subtle and subconscious opposition of the patient to the operator. Or again the open expression of sexual interest by a young child is clear evidence in favour of ''infantile sexuality'', while the absence of such expression is evidence of ''repression''. It is somewhat disconcerting to find that what is ostensibly a psychological hypothesis to be tested is in reality a faith to be embraced or rejected.' ([1913] p. 930).

There is no question of demonstrating Woodworth to be mistaken in the attitude he takes up but it is worth pointing out that the unfalsifiability of a psychoanalytic thesis may sometimes be due to a situation's being rich in possibilities rather than to the theory being rich in excuses. What the situation calls for is not a hard and fast rule but what Duhem calls 'good sense' (and, he might have added, good faith). ([1962] p. 218).

The difficulty in demonstrating, for example, that Freud's invocation of a wish to prove him wrong in order to account for 'counter-wish' dreams supports the description of him as a pseudo-scientist does not lie in doubt about the independent testability of such an auxiliary hypothesis but in doubt as to whether he employs it to justify his treating the apparently negative instance as a positive one instead of resting content to leave the matter *sub-judice* in the absence of further evidence. Consider the following specimen of psychoanalytic discourse: 'Delusions of grandeur ... are initiated by a withdrawal of libido from external objects which give rise to a pathological increase of narcissistic libido. The now excessive charges of libido in the ego stimulate feelings of grandiosity which may reach delusional intensity. This finding was subsequently corroborated by investigation of the traumatic neuroses of war.' (Glover [1956] p. 54). Isn't a difference made to the epistemic character of this specimen of discourse by that last sentence? It matters less whether we say that the exposition was merely non-scientific until we come to the remark reporting corroboration, and only then became pseudo-scientific, or that at that point one kind of pseudo-science became another kind of pseudo-science, than that the distinction be recognized.

In many discussions of the pseudo-scientific character of psychoanalysis the relevant distinction is not between falsifiable and non-falsifiable theories but between those which mislead expectation and those which merely under-determine it. It makes all the difference whether the content of a thesis was diminished *post hóc*, or whether it was candidly meagre from the outset. If it were indeed the case that 'psychoanalysis itself' was unfalsifiable 'from the start', this is a less discreditable state of affairs than obtains with regard to those components of Freud's theory of which this is not true.

Granted that a theory has been shown to contain qualifications which cancel or diminish its contents, why is this of any interest? Doesn't it only seem so because it is conflated with another question which is of paramount importance – whether the theory owes the impression of having been repeatedly confirmed to our overlooking the self-cancelling qualifications which are responsible for its resistance to falsification? That it is this which matters can be shown by asking what interest these self-cancelling qualifications would have if the tests that the theory survived did not involve them. For example, if the claims as to the reconstructive powers of psychoanalytic dream interpretation were not based on the invocation of ancestral memories to account for apparently erroneous reconstructions, would it matter that the notion of phylogenetic inheritance was a part of psychoanalytic 'theory-in-itself' and rendered erroneous reconstructions irrefutable? The question then becomes whether Freud was a pseudo-falsificationist with respect to his developmental theses, whether he issued what, in the light of the character of his concepts and methods, were spurious and unwarranted confirmation (in the sense of failed attempts at falsification) claims? There are reasons for thinking so. For example, 'My surprising discoveries as to the sexuality of children were made in the first instance through the analysis of adults. But later (from about 1908 onwards) it became possible to confirm them in the most satisfactory way and in every detail by direct observations upon children.' (Freud [1925] p. 70). If Escalona's account is correct Freud was perjuring himself.

What are Grünbaum's own criteria of pseudo-science? In several papers devoted to the subject it is remarkable that he should leave us in doubt as to his answer to this question. In an intellectual culture prolific in irrationalities Grünbaum can only think of two examples of pseudo-science, fatalism and Hartman's version of metapsychology (NB: not Freud's). He ignores Popper's own examples of astrology and the racial theory of history. He can't be oblivious to the defienciencies of a formalist reply (lack of P-falsifiability). On the other hand, it isn't certain that we are to equate Grünbaum's lack of 'I-scientificality' with pseudo-science, since he himself speaks of theories which fail blatantly to meet his standards of 'I-scientificality' ('I' for Inductive), not as pseudo-scientific but only as 'gratuitous' and 'lacking in scientific credibility'. But if we can take these idioms as equivalent to 'pseudo-scientific' it is possible to arrive at a piquant conclusion: Freudian psychoanalysis is pseudo-scientific by Adolph Grünbaum's own criterion of demarcation. He himself only avoids this conclusion through gross inconsistency. The expressions 'gratuitous' and 'lacking in scientific credibility' occur in his discussion of Erich Fromm's denial that research into psychotherapeutic efficacy, which has uncovered a tie-score effect between psychoanalytic and other modes of therapy, undermines the claims of psychoanalysis to be the treatment of choice for neuroses. Fromm's claim is stigmatized as 'gratuitous' because of his failure to provide evidence for the greater severity of the neurotic conditions of those treated by psychoanalysis. But if a thesis is 'gratuitous and lacking in scientific credibility' because of the failure of its advocates to take up the burden of proof then Grünbaum would have to say no less of Freud's developmental theory, since the advocates of the Freudian account of the infantile pathogenesis of the neuroses, when anticipating counter-examples, invoke a constitutional factor to account for neurotic outcomes of apparently non-pathogenic infancies without supplying independent evidence for the possession of such infantile constitutions, even acknowledging that the identification of such constitutions is not possible in our present state of knowledge.

Grünbaum has a second line of argument against Popper in addition to the denial of the unfalsifiability of psychoanalytic theory, though he doesn't always clearly distinguish it. He gives several examples of what he takes to be Freud's modifying his view in the face of contradictory evidence, of which the seduction theory episode and the case of the Rat Man are the most prominent. None of his examples support his thesis. This is what he says about the seduction episode: 'I am simply astonished that Popper did not see fit to try to deal with some of the hypotheses actually invoked by Freud to furnish explanations such as his famous infantile seduction theory of hysteria. I cannot help wondering, for example, whether Popper knew the following: Freud had hoped to make his reputation by solving the riddle of hysteria, but in 1897 he was painfully driven by unfavourable evidence to abandon his strongly cherished aetiological hypothesis.' ([1979b] p. 135).

Why does Grünbaum think that the seduction episode overthrows Popper's view of Freud as a dogmatist? In part because he is ill-informed as to the circumstances under which the seduction episode was embraced and then abandoned, and in part because he asks the wrong questions.

Here is how the seduction error struck another psychoanalyst (from a work which gets into Grünbaum's bibliography but seems to have made little impression on the way): 'A striking example of the essential irrefutability of psychoanalytic doctrine is afforded by Freud's handling of his discovery that his patients confabulated infantile memories, in itself strong evidence for the influencing value of his techniques. As he was quick to see, "this discovery ... serves either to discredit the analysis or to discredit the patients upon whose testimony the analysis, as well as the whole understanding of neurosis is built." This is a bleak predicament indeed, from which Freud extricates himself by a *tour de force*. He points out that "these fantasies possess psychological reality in contrast to physical reality" and "in the realm of neurosis the psychological reality is the determining factor." Therefore, the fact that these infantile experiences were fantasies rather than actualities, far from refuting his experiences, actually *confirms* them.' (Frank

[1961]). Why did this view not goad Grümbaum into a degree of self-consciousness as to the conclusiveness with which Freud's response to the seduction error illustrates his willingness to 'abandon a strongly cherished aetiological hypothesis'?

The question to be asked concerning the seduction theory is not just whether Freud manifested a willingness to consider evidence that hysterics need not have been sodomized as infants, nor even whether he was willing to consider evidence that those patients whom he had decided had been sodomized had not, after all, been sodomized, but whether he was willing to consider that he was in error in claiming that to the analyst, equipped with Freud's techniques of interpretation, the infantile past of the patient can be reconstructed through an analysis of his dreams, associations and behaviour during the analytic hour.

Grünbaum even misapplies Popper's criterion of *formal* falsifiability, for it entails asking not merely whether the seduction theory was falsifiable, but also the degree to which its successor theory was. Whether or not Freud's post-seduction infantile-sexual pathogenesis is falsifiable may be disputable but that it has less empirical content than the parent theory is not. It is universal where the parent theory was particular (and so is less testable epidemiologically)[5]. It has less determinate observational implications than the seduction theory (hidden fantasies for public events). And it was deprived of prophylactic implications. Whereas had the seduction theory been correct parents had only to keep a watchful eye on their children, or control their own peadophilic impulses, for the neuroses to have been eliminated in one generation, psychoanalytic pronouncements on the prophylactic implications of Freud's post seduction developmental theory are evasive and incoherent.

Grünbaum's argument is defective in another respect. For it is obvious that the capitulation to falsification reports can only reflect favourably on the integrity of the theorist where the refusal to capitulate would not undermine his credibility. But what choice had Freud? He could not say in the case of the seduction theory what his followers have said in the case of his theories of infantile sexual fantasy, that it took years of training to tell whether a child is being sodomized.

Grünbaum says both that Freud's patients had non-veridical memories of infantile seduction and that Freud based his seduction theory on these memories. Although this is Freud's own (later) account of the matter, it is false. There are several remarks in the original seduction papers with which it is inconsistent of which this is one:

> ... whilst calling these infantile experiences into consciousness they experience the most violent sensations, of which they are ashamed and which they endeavour to hide, and they still try, even after going through them again in so convincing a fashion, to withhold belief by emphasizing the fact that they have no feeling of recollecting these scenes as they had in the case of other forgotten material.
>
> Now this last attitude on their part seems absolutely decisive. Why should patients assure me so emphatically of their unbelief, if from any motive they had invented the very things that they wish me to discredit. ([1896] p. 204).

What this passage shows is that Freud based his conviction that his patients had been seduced not on his uncritically crediting their narratives of infantile seduction but on their behaviour being, in his judgement, 'consistent with no other assumption'. The motive for Freud's distortion is transparently clear from Jones' account of the episode: '[Freud] found that several of the seduction stories were simply untrue, there had been no seduction. But he held fast to the fact that *the patient had told him these stories* ... with the result that he discovered the importance of infantile fantasy life in the genesis of the neuroses.' [My italics]. ([1939] p. 111).

Grünbaum's account of Freud's *abandonment* of the seduction error is also erroneous. He tells us that one of Freud's 'explicit reasons' for abandoning the seduction theory was his discovery of the unexpectedly high incidence of hysteria which, since seduction was only a necessary condition of hysteria, would have involved him in postulating a 'preposterously high' incidence of child-seduction. But since Freud tells us that he discovered that some of his imputed seductions could not have occurred, that is, not all 18 of his patients had been seduced, he could not have abandoned the seduction theory 'for several explicit reasons', as Grünbaum asserts, since the discovery that some of the 18 had not been seduced rendered any additional reasons otiose.

As for the error of mistakenly attributing seductions to several of his patients, this is Grünbaum's account of how the hypothesized seductions were (erroneously) 'confirmed': 'The mentioned *prima facie* confirmations of the postulated seduction episodes had been furnished by the seemingly vivid and presumably repressed memories that Freud had been able to elicit from his hysterical patients in the course of their analysis.' In the same paragraph Grünbaum speaks of the 'subjective certainty felt by his adult patients in the reality of purported memories going back to childhood'. ([1979b] pp. 135 – 6). (And we are referred to the letter in which Freud repudiated the seduction theory).

The passage I quoted is difficult to reconcile with Grünbaum's claim that Freud was the recipient of false seduction reminiscences, or that patients felt 'subjective certainty' as to non-veridical infantile seductions. His mistake may spring from reliance on secondary sources for his account of the seduction episode. (The seduction papers themselves do not figure in his bibliography.) The mistake is a serious one for it prevents Grünbaum from realizing that Freud's response to the seduction error, far from illustrating his willingness to abandon the erroneous theory, provides some of the strongest evidence of his inability to take up a critical attitude towards his own convictions; an inability which caused him to give a tendentiously distorted account of how he came to make the seduction error and thus relieved him of the painful necessity of confronting the real implications of his mistake. (Cioffi, [1976] pp. 275 – 80).

4. *Does the Rat Man case illustrate a 'testing strategy'?* As a further counter-example to Popper's 'indictment of Freud' Grünbaum refers us to a paper by Clark Glymour which advances the argument that Freud's response to the facts of the case history of the Rat Man demonstrates that in his aetiological speculations and investigations Freud was pursuing 'a testing strategy': 'The best available evidence concerning the actual life history of his "Rat Man", Paul Lorenz, had *refuted* his prior hypothesis as to the specifics of the sexual aetiology which he had postulated for adult obsessional neuroses. (cf. Glymour [1974] pp. 299 – 304).' ([1979b] p. 137). There was no 'prior hypothesis as to the specifics of the sexual aetiology ... for adult obsessional neuroses'; and, if there had been, 'the best available evidence concerning the actual life history' of the Rat Man could not have refuted it; and if it had, and Freud had responded as Glymour describes, it would not have illustrated Grünbaum's claim that Freud manifested a readiness to abandon his theses if they conflicted with the evidence.

It would be difficult for any infantile life to contradict the views Freud held about the aetiology of the neuroses after his abandoning the seduction theory, and even more difficult to discover whether it, in fact, had. This is Freud's first post-seduction statement of the infantile events which predispose to neurotic disorders: 'Their real aetiology is to be found in experiences during childhood, and again exclusively, too, in those impressions which have to do with sexual matters.' ([1898] p. 242). It is difficult to see how the Rat Man's case history, or any case history for that matter, would reveal that the patient's infancy contained no 'impressions which have to do with sexual matters'.

In a paper of 1905 ('My Views on Sexuality in the Etiology of Neuroses') Freud said of the neuroses, 'these maladies are observed to ensue after the most commonplace emotional, or even somatic, disturbances and a ... specific aetiology in the form of a particular infantile experience is not forthcoming'. ([1905a] p. 281) In the *Three Essays on the Theory of Sexuality*, also 1905, Freud says: 'Of the various factors which interfere with development, first and foremost, we must name the innate variety of sexual constitutions, upon which it is probable that the principal weight falls.' (The summary to *Three Essays on the Theory of Sexuality*, paragraph entitled 'Constitution and Heredity'). In the seventh paragraph of this Summary he says, 'it was not possible to say what amount of sexual activity can occur in childhood without being described as abnormal or detrimental to further developments'.

Glymour maintains that Freud introduced infantile sexual fantasies into the aetiology of the neuroses because he failed to find in the infantile life of the Rat Man the sexual events to whose pathogenic potency he was up till then committed: '... before (the Rat Man case) there is no statement of the view that sexual fantasies formed in childhood or subsequently, having no real basis in fact, may themselves serve in place of sexual experiences as aetiological factors.' ([1974] p. 303); 'increasingly after 1909, Freud thought fantasies themselves, even when derived from no real sexual experience, could serve as aetiological factors.' ([1974] p. 302)

The distinction between fantasies 'derived from real sexual experiences' and fantasies not so derived needs to be more clearly drawn before we can say with any confidence that only after a certain date did Freud replace or supplement one with the other. In the Dora case (1900) a pathogenic role is assigned to fantasies derived from her experience of thumb-sucking and of the sight of a cow's udder. If these are not sexual experiences in Glymour's sense, and Dora's fantasies 'derived from no real sexual experiences' then the Dora case is a counter-example to his chronological claim; if, on the other hand, they do qualify as sexual experiences then Freud could not have found that there were no experiences of this character in the Rat Man's childhood, or in any other. In any case, how could Freud discover that the infantile sexual fantasies of the patient 'derived from no real sexual experience' when Freud considered that even babies being suckled and washed occasion sexual experiences? ([1905a] pp. 136–7)

Even had Freud held the view of infantile sexual pathogenesis that Glymour attributes to him, i.e. that sexual intimidation was a necessary condition of obsessional neuroses, the facts in the case of the Rat Man could not have compelled him to modify them and substitute fantasies for events as aetiological factors. Glymour gives the following reasons for believing that 'the clinical evidence in the Rat Man case tested views Freud held': 'That the inferred castigation for sexual misbehaviour was not revealed by the most reliable means – the memory of an adult observer – is presumably evidence that something was wrong with Freud's account of the role of psychosexual development in the aetiology of the neuroses.' ([1974] p. 301)

The germ from which Glymour's argument seems to have grown is the footnote in the case history of the Rat Man in which Freud records his disappointment at learning that the offence for which the Rat Man had been beaten was not connected with onanism, as Freud had conjectured. Though the Rat Man *had* been beaten, enquiries Freud asked him to make of his mother revealed that it had not been for a sexual offence. Instead of insisting that the mother was mistaken, Freud graciously condescends to leave the matter open and Glymour credits him with a falsificationist methodology on the strength of it.

In asserting that the Rat Man had been punished by his father for masturbating Freud was not evincing his commitment to a previously stated law linking a specific infantile history with the symptom picture presented by the Rat Man. In the first place because he was committed to no such law and in the second, because the 'non-theoretical state of affairs, namely, the patient's having been punished by his father for masturbation' was not *inferred* by Freud. He was only *repeating* his patient's story of having been beaten, and *adding* the suggestion that it had been for a sexual offence. The passage from the case history, to the effect that Freud had anticipated the beating episode, which Glymour quotes, was doctored (or whatever euphemism Freudians prefer) after Freud had discovered that the Rat Man had been 'castigated' as a child. If Glymour had read the 'Original Record' with a modicum of care he would have discovered that the original version of the passage he quotes contains no allusion to Freud's anticipation of a 'castigation' suffered by the Rat Man, which was then confirmed by family tradition, thus furnishing evidence of Freud's uncanny prescience. Freud merely conjectures, erroneously as it turns out, that the Rat Man had been *warned* as to his masturbatory practices and 'perhaps' threatened with castration.

This is what Freud says in the published case history:

> Starting from these indications and from other data of a similar kind I ventured to put forward a construction to the effect that when he was a child of under six he had been guilty of some sexual misdemeanour connected with onanism and had been soundly castigated for it by his father. This punishment, according to my hypothesis, had, it was true, put an end to his onanism, but on the other hand it had left behind an ineradicable grudge against his father and had established him for all time in his role of an interferer with the patient's sexual enjoyment. ([1909] p. 342)

This is what he says in the notes he made at the time:

> I could not restrain myself here from constructing the material at our disposal into an event: how before the age of six he had been in the habit of masturbating and how his father had forbidden it, using as a threat the phrase, 'it would be the death of you' and perhaps also threatening to cut off his penis. ([1909] p. 263)

Notice the 'perhaps'. Freud did not even feel himself aetiologically committed to castration *threats*, much less the castigation for sexual misbehaviour which Glymour attributes to him.

Not only does the failure of the Rat Man's interrogation of his mother to confirm Freud's conjecture not conflict with 'Freud's account of the role of psychosexual development in the aetiology of the neuroses', it does not even provide a test of Glymour's own misconception of this. It merely rendered less plausible one *particular* reconstruction that Freud had ventured in the course of the treatment. Why should Freud's disappointment over the reconstruction of one particular episode have compelled him to abandon his view that the Rat Man had been sexually intimidated? How do the facts recorded in the footnote show that the Rat Man had not been sexually intimidated by his father on some other occasion? In fact we know that he was later beaten for bedwetting (which Freud held to be a surrogate sexual activity). Freud's insistence that the Rat Man was only beaten once, which Glymour repeats, is a slip. (Original Record [1909] p. 384)[6].

Even supposing Freud had held the views that Glymour attributes to him and that the facts in the Rat Man case had contradicted those views, would Freud's behaving as Glymour describes and invoking the pathogenic power of sexual fantasy exemplify a testing strategy? Of what must we be persuaded if we are to concede that Freud's introduction of aetiological factors, which lie in the fantasy life of the child and not in publicly observable circumstances, manifests the willingness to abandon theses inconsistent with the evidence with which Grünbaum and Glymour credit him? How are we to distinguish between Freud's successor thesis to the seduction theory and that of the Christian Scientist who evokes Malicious Animal Magnetism to account for the fact that even Christian Scientists succumb to illness, or the astrologer who explains his faulty forecasts by invoking the influence of the sky-pattern at conception to supplement that at birth? Presumably we do not wish to say that these, too, were evincing their loyalty to a testing strategy and their willingness to abandon theories incompatible with the evidence. In order for us to acknowledge a difference between these cases and Freud's we would need to be persuaded that in Freud's case there is some hope of discovering both what the fantasy life of the child is, with sufficient determinacy to distinguish between rival accounts of its nature, and also how to distinguish its influence on the character of the child from those of other eligible determinants. At present we can do neither of these. Whether this may have been among Freud's reasons for adopting the fantasy theory we will, on this occasion, leave to heaven. If any one can be said to have employed 'a testing strategy' with respect to Freud's infantile sexual vicissitude theory it is those erstwhile Freudians (from Ian Suttie to John Bowlby) who, on failing to find differentiating sexual vicissitudes in the infantile pasts of their patients (as Freud himself had), instead of abandoning the search for personality-malforming influences in the overt life of the child and retreating to fantasy, sought for it in the interpersonal vicissitudes of child-life and ceased demanding that these be sexual. It may redound to Mary Baker Eddy's credit that she abandoned homeopathy on discovering the operation of a placebo effect but does it redound to her credit that she replaced it with Christian Science? (cf. Cioffi [1970] p. 473).

Grünbaum is inconsistent in invoking Glymour's paper to confute Popper since its argument moreover violates the requirement he employs elsewhere, e.g. in his treatment of the claim that in spite of the lower remission rate for psychoanalytically treated patients psychoanalysis may, nevertheless, be more effective since its patients may be sicker. Grünbaum demands independent evidence that this is the case and since it is not forthcoming stigmatises the claim as 'gratuitous and devoid of scientific credibility'. If the same considerations were applied to the invocation of the pathogenic role of infantile fantasies the same conclusion would follow, for where is the independant evidence that those who later succumb to neuroses are characterised by any distinctiveness in their infantile sexual fantasies. In fact Freud conceded that there was none.

Grünbaum has also argued that 'the conjecture of the sexual aetiology of phobias makes some predictions that are "risky" by Popper's standards', for example, that a non-psychoanalytic method of removing phobic symptoms would result in symptom-substitution. 'But contrary to this psychoanalytic prediction a 1978 report from the Institute of Psychiatry at the University of London concludes: in the overwhelming number of cases the reduction or elimination of a circumscribed fear is not followed by untoward effects. So much for Popper.' ([1979a] p. 514) This is a strange argument. If it is meant to demonstrate that the psychoanalytic theory of phobias has observable implications it would be sufficient to state these and point out their risky character without documenting the actual occurrence of the falsifying observations. That is, we wouldn't find it necessary, in order to demonstrate the falsifiability of 'all swans are white', to supply evidence of the existence of black swans. And so Grünbaum may be making a different point but one which leaves the argument equally strange: the willingness in practice of some investigators to consider the psychoanalytic theory of phobias falsified by certain observations demonstrates the non-pseudo-scientific character of the theory. But for this argument to bear on the *bona fide* non-pseudo-scientific character of psychoanalytic theory, the willingness in question ought to be on the part of those who hold the theory not, as in the case with Rachman, those who are among its severest critics.

In any case does capitulation to counter-examples suffice to establish the *bona fides* of either a theory or an investigator? An eminent contemporary astrologer reports in the introduction to his book that he had attempted to confirm a time-honoured astrological thesis and failed. 'In making lists of persons with two bodies in good aspect and of other persons with the same bodies in bad aspect we could find no difference in fortune.' Is this sufficient to show either that astrology is scientific, or that he himself is a *bona fide* scientist? It can show neither. It is not the status of the general claims alone which is pertinent to a discussion of the suspicion of pseudo-science but of the reporting practices with respect to particular existential claims as well. The mistake of assuming otherwise is made by Grünbaum when he argues that Freud's theory of melancholia is in good order since it is falsified by the existence of endogenous non-reactive depressions. It does not settle the matter to show that there are exceptions to Freud's explanation of melancholia, or even that they are conceded to be exceptions. What needs to be shown is that the identification of positive instances of the thesis meets acceptable standards.

A theorist may manifest a readiness to abandon his general claims in the face of apparent falsification and still be a pseudo-scientist. Since Grünbaum examines only Freud's general claims he is precluded from dealing with the central issue.

Grünbaum takes exception to the following remarks of Popper's:

> ... how much headway has been made in investigating the question of the extent to which the (conscious or unconscious) expectations and theories held by the analyst influence the 'clinical responses' of the patient? (I say nothing of the conscious attempts to influence the patient by proposing interpretations to him, etc.). Years ago, I introduced the term 'oedipus effect' to describe the influence of a theory, or expectation, or prediction, upon the event which it predicts or describes: it will be remembered that the causal chain leading to Oedipus' parricide was started by the oracle's prediction of this event. This is a characteristic and recurrent theme of such myths, but one which seems to have failed to attract the interest of the analysts; perhaps now quite accidentally. ([1969] p. 38).

Whereas 'some recognized analysts have dealt with troublesome considerations in a puzzlingly inconsistent and even evasive manner', Grünbaum thinks Freud did not, and Popper is charged with 'compounding his exegetical felony' (of suggesting that Freudians had not taken proper account of the contamination issue) by hinting that 'analysts are evasive or even repressing an uncomfortable challenge.' Grünbaum rebuts this presumed slander by invoking 'Freud's Tally Argument of 1917' which, though it may be 'unsatisfactory', absolves Freud from the charge of 'having offered a mere *ipse dixit*.'

This is Grünbaum's statement of the tally argument: 'Only psychoanalytic interpretations that "tally with what is real" in the patient can mediate veridical insight and such insight in turn, is causally necessary for the successful alleviation of the patient's neurosis.' ([1979a] p. 467). Grünbaum dates the tally argument from the *Introductory Lectures* of 1916 – 17, but Freud was implicitly relying on it a quarter of a century earlier in the *Studies on Hysteria*. Since the seduction error intervened between this work and Freud's restatement of the tally argument in 1917 why shouldn't Popper be surprised (not 'dumbfounded' as Grünbaum tendentiously puts it) at Freud's assertion of the harmlessness of suggestion? In any case how could Freud advance a therapeutic argument for veridicality when once he had the seduction error behind him and thus knew that he could achieve effects without veridicality?

There are reasons for thinking that Freud himself was not entirely convinced by the tally argument. He sometimes implied its invalidity. In 1910 he wrote: '... I need not rebut the objection that the way in which we practise the method today obscures its testimony to the correctness of our hypothesis; you will not forget that this evidence is to be found elsewhere and that a therapeutic procedure cannot be performed in the same way as a theoretical investigation.' ([1910] p. 287)

So Freud did not always invoke the tally argument to support the validity of his method. Instead he appealed to the fact that psychotics express thoughts and wishes of a similar character to those he uncovered by interpreting symptoms. Though this might, if one were inclined to generosity, be taken as evidence that neurotics harbour the same wishes that are explicitly expressed by psychotics how could it show that it is these wishes which are responsible for the neurotic symptomatology Freud invoked them to explain? How would a psychotic's direct expression of incestuous wishes assuage our doubts as to whether the bedtime ritual girl's transactions with her pillow and bolster was an expression of her wish to be penetrated by her father? How does the fact that a female psychotic expresses blatant defloration fantasies help to show that Freud was correct in assigning an hysterical patient's headache to this source?

Freud also seems anxious, to the point of prevarication, to deny that he held his convictions as to the sexual source of neuroses before the material which confirmed them was forthcoming. If Freud thought the operation of suggestion 'epistemically innocuous', as Grünbaum maintains, why did he so often deny (untruthfully) that he had held his convictions as to the sources of neurotic illness *before* he received the patient's confirmation of them? In 'The Question of Lay Analysis' he wrote: 'The analyst never entices his patient on to the ground of sex. He does not say to him in advance: "We shall be dealing with the intimacies of your sexual life!" He allows him to begin what he has to say wherever he pleases, and quietly waits until the patient himself touches on sexual things. I used always to warn my pupils: "Our opponents have told us that we shall come upon cases in which the factor of sex plays no part. Let us be careful not to introduce it into our analyses and so spoil our chance of finding such a case." But so far none of us has had that good fortune.' ([1925] pp. 118 – 19). And yet in the very paper on dream interpretation, to Popper's comments on which Grünbaum objects, Freud speaks of patients who 'reproduce the forgotten experience of their childhood only after one has constructed them from their symptoms, associations and other signs and has propounded these constructions to them ... unless one interprets, constructs and propounds one never obtains access to what is repressed in them.' ([1923] p. 142)

Here is another illustration of Freud's tendency to back away from the tally argument and inadvertent testimony to the force he really attributed to Popper's Oedipus effect – Freud's false retrospective accounts of the state of his theory at the time he undertook the analyses which led him into the seduction error. In the autobiographical sketch of 1925 he wrote: 'It would have been difficult to guess from the *Studien über Hysterie* what an importance sexuality has in the aetiology of the neuroses.' ([1925] p. 39). And yet in the preface to that very book Freud and Breuer wrote: 'Sexuality plays the principal role in the pathogenesis of hysteria as a source of psychic traumas, and as a motive of "defence" and of the repression of ideas from consciousness.' ([1895] p. xxx). And in one of the seduction papers themselves Freud refers to the *Studies in Hysteria* in terms which acknowledge his prior committment to a sexual aetiology in that work: 'In earlier publications Breuer and I have already expressed the opinion that the symptoms of hysteria can be understood only by tracing them back to traumatic experiences and that these psychical traumas are related to the patient's sexual life.' ([1896] p. 156).

It becomes a mystery why, if Freud really believed the tally argument rendered the operation of suggestion 'epistemically innocuous', he should have gone to such lengths (whether consciously or not) to deprive the Oedipus effect argument of force by rewriting the oracle's prophecies.

It is a measure of Grünbaum's obsequiousness where Freud is concerned that he can consider something as natural as the suspicion that Freud was behaving evasively in his dealings with the suggestion argument as a slander. The feeblest of rationalizations counts in Grünbaum's eyes as 'facing the problem of suggestion'.

5. *Conclusion* I have given my reasons for believing that it is a mistake to maintain that 'Popper's own criterion of demarcation does not entitle him to indict psychoanalysis as unscientific ...' (Grünbaum, [1976] p. 216). I am not sure whether 'falsifiability is the touchstone of scientific rationality', but what Grünbaum calls Popper's indictment of Freud does not depend on its being so.

The idea that the mere possession of potential falsifiers shows that a theory is not pseudo-scientific, which Grünbaum mistakenly assigns to Popper, and sometimes seems to hold himself, is wrong. Grünbaum's idea that whether a theory has falsifiers can always be determined independently of the behaviour of its advocates is also wrong; and so is his idea that the empirical content of a theory can be determined by consulting its critics.

Grünbaum's most fundamental (non-exegetical) error is that, when he has succeeded in demonstrating that a thesis is formally falsifiable and thus scientific in a sense antithetical to non-scientific or metaphysical, he thinks he has shown that it is scientific in the sense in which the term is antithetical to pseudo-scientific. Grünbaum slides from 'the tenacity of its adherents cannot show that a theory is unfalsifiable' to 'the tenacity of its adherents cannot show that a theory is pseudo-scientific', from which it would follow that one of the standard uses of the term pseudo-scientific is a solecism.

Furthermore, what Freud presents us with is not merely a theory but copious specimens of his explanatory practice, and these are just as capable of providing an indictment of psychoanalysis as a pseudo-science as the wanton rejection of falsifications of its general claims. Grünbaum, who fails to address himself to the justice or otherwise of the complaint that its singular instantiation reports are unwarranted, is disqualified from making a substantial contribution to the discussion.

Once we recognize that Karl Popper's criterion of demarcation is not merely formal falsifiability but 'severe attempts at refutation', we will give an affirmative answer to Grünbaum's question – Freudian psychoanalytic theory *is* pseudo-scientific by Karl Popper's criterion of demarcation.

Notes

1. There is another ambiguity worth mentioning: whether an auxiliary hypothesis, introduced to save a theory from falsification, must, if it is not to count as *ad hoc*, merely entail *entertainable* (conceivable) falsifiers, or whether these must be *ascertainable* as well. (This is akin to Carnap's distinction between testability and confirmability). Popper speaks at one point of the 'at the time untestable' character of *ad hoc* hypotheses. This sounds like an objection to their lack of ascertainable, rather than of entertainable, falsifiers. One of Popper's expositors seems to equate falsifiability with the possession of ascertainable and not merely entertainable observation statements. When J.O. Wisdom states that had there been no moon, or no planet Mercury, Einstein's general theory would have been less falsifiable, it is not the notion of entertainable falsifiers that he is employing but that of ascertainable ones, since the logical implications of Einstein's theory would have been the same even had there been no moon and no planet Mercury. That is, if the day before the

solar eclipse of 1919 the moon had exploded, or gone off into space, Einstein's theory would have become less testable but not less confirmable, or, in my terms, it would have had fewer ascertainable falsifiers but no fewer entertainable ones. The issuance of confirmation reports where theories lack ascertainable falsifiers is as disreputable as it would be had they lacked entertainable ones.

2. In 'Is Falsifiability the Touchstone of Scientific Rationality?' ([1976] p. 228) Grünbaum says, 'For Popper the mere falsifiability of a hypothesis for according scientific status to it, but the inductivist may be prepared to grant it no more than potential scientific status in virtue of its inductive supportability.' But the sense of 'scientific status' in which Popper allows it is not the same as the sense in which the inductivist denies it. For Popper the demonstration that a thesis has falsifiers may be a defence against its characterization as metaphysics but not against its characterization as pseudo-science. (My objections to Grünbaum's dealings with the relative stringency of inductivist and falsificationist criteria of pseudo-science is completely independent of the cogency or otherwise of his defence of inductivism.) There is even some doubt whether formal falsifiability is considered by Popper to be inconsistent with the metaphysical ('non-scientific') status of a speculation. At one point he gives the 'readiness to look out for tests' as distinguishing science not merely from pseudo-science but 'especially from pre-scientific myths and metaphysics'. Aristarchus is said not to qualify as a scientist yet his cosmic heliocentrism is said to be false and his local heliocentrism to have been confirmed by Kepler, which seems inconsistent with either lacking falsifiers.

3. This is the weakest of my exegetical reconstructions of Popper. Though he does concede that a certain degree of dogmatism is required for science to advance, and that there are occasions when the rejection of putative falsifications is justified ([1968a] p. 81), he also sometimes speaks as if the resort to *ad hoc* hypotheses is sufficient to brand someone a pseudo-scientist. Whichever view we assign him on this issue it would not affect his adverse account of psychoanalysis since, though unequivocal refutations of its theories are not common and so their tenacity can be justified, this could not extenuate the unwarranted and spurious confirmation reports which are a staple of the literature and which is the alternative basis of Popper's indictment of it. Even if Popper does sometimes suggest that unfalsifiable theses are on that account alone pseudo-scientific, he is then using 'pseudo-scientific' in a different sense than that in which he employs it when he complains about incessant and unwarranted confirmation reports ('the stream of verifications'). Here is another remark which shows Popper objecting neither to the unfalsifiability of a theorist's claims, nor to his refusal to capitulate to apparent falsifications but to his reporting practices: 'The most characteristic element in the situation seemed to be the incessant stream of confirmations, of observations which verified the theory in question; and this was a point constantly emphasized by their adherents ... The Freudian analysts emphasized that their theories were daily, nay, hourly verified by their "clinical observations" '.

4. From time to time Freud himself conceded the untestability of his dream theory. In the *New Introductory Lectures* chapter on dreams he said that though every dream was meaningful not every dream could be interpreted. In the *Outline* he said, 'It is often hard to detect the unconscious motive force and its wish-fulfilment but we may assume it is always there.' ([1904a] p. 33)

5. Freud invoked a constitutional factor to explain why universal features of infantile life, like incestuous fantasies, should nevertheless only sometimes be pathogenic but objected to the emptiness of this explanation when invoked by others. In his *Mental Hygiene* review of *The Problem of Anxiety* Rank chides Freud for his inconsistently failing to concede that the constitutional factor might just as readily be invoked to meet Freud's own objections to Rank's birth trauma theory. (Rank [1927] pp. 185 – 188)

6. This is Freud's account of the second beating in the 'Original Record': 'When he was a little boy (age uncertain, perhaps five or six) he was lying between his father and mother and wetted the bed, upon which his father beat him and turned him out.' ([1909] p. 384)

Glymour commits another exegetical error: 'Freud was already committed to the primacy of masturbation as a form of sexual gratification in infancy and early childhood and as an aetiological factor in neurosis. Indeed he repeats this commitment in the very same section of the case study in which the previous quotation was taken.' Glymour then quotes in support: 'Infantile masturbation reaches a kind of climax, as a rule, between the ages of three and four or five, and it is the clearest expression of a child's sexual constitution, in which the aetiology of subsequent neuroses must be sought.' ([1909] p. 202) On this occasion Glymour is the victim of a genuine ambiguity in Freud's text rather than his own tendentiousness. It is clear from the German version that it is to the sexual constitution and not masturbation that we are to look for the etiology of subsequent neuroses. If avoiding this ambiguity were the only consideration the translation ought to run: 'Masturbation ... is after all the plainest expression of the sexual constitution of the child, to which constitution we must look for the aetiology of the neuroses.' ('Aber die Onanie der Pubertätsjahre ist in Wirklichkeit nichts anderes als die Auffrischung der bisher stets vernachlässigten Onanie der Kinderjahre, welche zumeist in den Jahren von 3 bis 4 oder 5 eine Art von Höhepunkt erreicht, und diese ist allerdings der deutlichste Ausdruck der sexuellen Konstitution des Kindes, in welcher auch wir die Aetiologie der späteren Neurosen suchen.' ([1909] pp. 423 – 4).

3. POPPER AND THE MIND-BODY PROBLEM

Roland Puccetti

As so often in his career, Karl R. Popper has swum against the tide on the mind-body problem too. In *The Self and Its Brain* [1977], co-authored with John C. Eccles, Sir Karl gives us what is probably his last and certainly his most extensive word on the subject. Here I shall concentrate on just those features of the account which I take to be crucial and of enduring interest to philosophers and brain scientists alike.

1. *The grand reductio*. In Chapter P3 of the book entitled *Materialism Criticized* Popper invokes evolutionary considerations against the now largely defunct theory of epiphenomenalism. In brief, if mental events are causal by-products of brain processes, but themselves have no effects, how can one possibly explain their emergence in animal evolution? To use Popperian terminology, no-one would seriously question that World 1, the physical world, antedates World 2, the world of mental occurrences. If so, and if one does not want to postulate divine intervention but rather take a Darwinian view of things, then epiphenomenalism must be rejected because it denies any biological function to a plausibly widespread (among higher animals) natural phenomenon[1].

Now it has often been asserted that the theory known as psychophysical identity cannot succumb to this line of criticism, since if a given mental event or kind of mental event is assumed to be *identical* with a particular brain process, then whatever adaptive advantage was accrued by evolving the latter is conferred by the former as well: they are one and the same. It is to Sir Karl's credit, I believe, that he sees through this manoeuvre. For in Section 23 of the same chapter, specifically titled 'Does the Identity Theory Escape the Fate of Epiphenomenalism?', he observes that the identity theorist would reduce

Currie, G. and Musgrave, A. (eds.), Popper and the Human Sciences.
 ISBN 90-247-2998-X

World 2 to a subclass of World 1, which he labels World 1_m: all other World 1 (there being no World 2 or 3) occurrences being designatable as World 1_p events. By this reductionist move the motivation for physicalism is preserved, i.e. what Popper characterizes as the *causal closedness of the physical world*. In this respect epiphenomenalism and the identity theory are exactly alike, for they both deny that distinctly mental events have any causal role in nature: epiphenomenalism because they are causeless by-products of brain processes, psychophysical identity because there are no distinctly mental events[2].

Popper concedes that the identity theory is intuitively more satisfying than epiphenomenalism, since it allows for causal interaction between World 1_m (mental-physical) events and World 1_p (purely physical) events in the rest of the nervous system and the extracorporeal environment. Yet he also sees that his Darwinian critical stance carries over to strike equally at the heart of the identity theory.

> For the identity theory is, by intention, a purely physicalistic theory. Its fundamental principle is still the principle of the closedness of World 1; which leads to the lemma that (causal) explanation, so far as it is knowledge by description, must be in terms of strictly physical theory. This allows us (perhaps) to accept the emergence of a new World 1_m; *but it does not allow us to explain that the characterizing feature of this World 1_m is that it consists of mental processes, or that it is closely linked with mental processes* (Popper and Eccles [1977] p. 87; author's italics).

Popper's insight here can be elaborated in the following manner. Suppose we were visited by advanced extraterrestrial physicalists who rapidly gain near-complete knowledge of our evolutionary history (and that of other higher animal forms on Earth), our neurophysiology, etc. They may even have sensory systems attuned to the discharging of our neurons. Now if epiphenomenalism were true, what could ever lead them to suspect that we have mental occurrences? Or, if psychophysical identity theory were true, how could they come to learn that some of the neural events going on in us are mental as well as physical events? Perhaps they could come to know this by studying our language and finding that we have mental terms that we learn to use without (unlike themselves) knowing what is going on in our nervous systems (knowledge by acquaintance). But this would not change the point. For the point is, if Popper is right, that *as physicalists*, nothing in our evolutionary past would explain this fact about us. Or to put it another way, by ruling out any distinctive causal role for mental events, physicalism cannot within its own terms of reference account for the emergence of the mental.

> The principle of the closedness of World 1 demands that we still *explain*, truthfully, my going to the dentist in purely physical terms. But if so, the fact that World $1m$ is identical with World 2 – the world of my pains, my aim to get rid of them, and my knowledge about the dentist – remains causally redundant. And this is not changed by the assertion that another causal explanation, a World 2 one, is also true: it is not needed; the world works without it. But Darwinism explains the emergence of things or processes only if they make a difference. The identity theory adds a new aspect to the closed physical world, but it cannot explain that this aspect is of advantage in the struggles and pressures of World 1. For it can explain this only if the purely physical World 1 contains these advantages. But if so, then World 2 is redundant (Popper and Eccles [1977] p. 88).

The argument can be made still stronger. Suppose our fancied extraterrestrial physicalists could see us transparently at any level they pleased, as if we were medical textbook models with layers of coloured flimsies one on top of the other. Then they might be able to regard us as self-contained physical systems receiving stimuli at the surface of our bodies, processing the information complexly, and reacting with appropriate behaviour. Indeed, many Earthling physicalists today talk as if this is how we work. But then clearly it should make no difference to our workings *which*, if any, of our component internal processes belongs to World 1_m and which to World 1_p. That some elements of the self-contained system should be both mental and physical could hardly add to their causal efficacy if, in the celebrated phrase, 'sensations are nothing over and above brain processes'[3]. But neither could it *detract* from those workings, in that case, should some components now taken to be both physical and mental become physical only. If mental-physical identity makes no causal difference to behaviour because we are basically nothing but physical systems, non-identity would make no causal difference either. So the cerebral cortex could, physicalism would have it, function just as well unconsciously and, as I suggested elsewhere, the cerebellum function consciously[4]. But this is grossly false, as far as we can tell, on sheer empirical grounds.

The upshot of Popper's *reductio* against epiphenomenalism thus spills over into the identity theory, with telling force. Extraterrestrial physicalists simply could not predict our behaviour (beyond reflexive movements mediated by spinal arc pathways or vestibular mechanisms) by looking within us. There would be gaps in the physical causal chains, gaps where something distinctly *mental* is going on that is part and parcel of the total causal nexus. But what exactly is this, and how does it manage to *interact* with the brain?

2. *The cortical module theory*. The precise account of mind-brain interactionism is provided not by Popper (though he apparently concurs in it), but by his co-author, John C. Eccles. In Chapter E7, 'The Self-Conscious Mind and the Brain', Sir John starts with the widely regarded columnar hypothesis of cerebral cortical organization. As I pointed out in an earlier review of the book, he likens the 'self-conscious mind' to a searchlight or scanning device that seeks out, or probes, and selects from columns of neurons on the surface of the 'liaison brain' (the left cerebral hemisphere in almost all of us, where speech and language comprehension are preponderantly represented)[5]. There in certain 'liaison areas' where there is linguistic or ideational representation, and in polymodal sensory areas, it finds open, or partly so, modules where it 'gets experiences'; it can even, if required, cross through the corpus callosum (the main fibre bundle connecting the two cerebral hemispheres at cortical level) to gain access to items of musical or pictorial interest in the right half brain. In dreamless sleep, coma or convulsions all the surface modules are closed, forcing the self-conscious mind to wait until they reopen.

These are metaphors, of course. But what possible neurophysiological relevance do they have? Assuming the columnar hypothesis to be well founded, there is no sense to speaking of neurons, individually or in columns, being open, partly or completely closed. Neurons either fire or they do not; transmitter substances cross the synaptic cleft and provoke or inhibit neuronal discharge, and that is basically the whole story. Apart from speaking vaguely about open modules having a certain level of dynamic activity in them, Eccles does nothing to justify these metaphors.

And certainly we cannot take them literally. For example, I am a self-conscious mind by Sir John's lights, yet I have never caught myself looking at neocortical columns in my brain. Could I be doing this *un*consciously? But seeking, probing and selecting are things I do consciously. And if I really could select experiences from the modules, why do I get toothaches and drenchings when I could, apparently, pass over those modules and select from less disagreeable ones? My brain is hardly a garden where I can pick and choose at will.

Indeed such loose language in modelling mind-brain interaction is misleading, for it makes it sound as if the mind could, if it wanted, *scan a different brain* and select experiences from *its* open or partly open modules (whatever that exactly means). But with human beings, and in this life at any rate (to avoid begging religious questions), the living brain is the anchor of personal identity. I cannot remember, really remember, someone else's childhood as my own, nor he mine. Embracing dualistic interactionism because it alone among the major theories provides for the causal efficacy of the mental must not be allowed to obscure the fact, gained through one hundred and fifty years' experience in clinical psychopathology, that the mind is completely dependent upon the brain. This is most dramatically revealed in selective brain damage, or therapeutic brain surgery. For example, the split-brain patient who has had all neocortical nerve connections severed, for relief of epilepsy, will if righthanded name all the objects to the right of fixation but none to the left, and is unaware that anything is missing from his visual field[6].

What, then, is the proper model of mind-brain interaction? First we must set aside two stock objections to the very notion of a causal relation between brain and mind.

Popper, in Chapter P5, Section 48 of *The Self and Its Brain*, deals effectively with the claim that no such relation is possible because, on dualistic assumptions, mind and brain have no common properties. He points out that this objection arose from the fact that Descartes was a believer in both mechanism (including, of course, the human brain) *and* in the doctrine of the incorporeality of the soul, at a time when physical causation was thought to be the result of mechanical *push*. But how could something essentially unextended, the mind or soul, push against something essentially extended, the inner surface of the pineal body (which, though not part of the brain, is at its base), and vice-versa? However, the 'mechanical push' view of physical causation has been dead since Newton formulated his laws (it cannot account, for example, for gravitational forces). Modern views of causation, Popper might have added, do not even require that we *understand* how one physical event causes another, provided that our equations have the requisite predictive power, as is clear from the current situation in quantum mechanics.

> The philosopher's question, 'But surely the uncertainty relations do not entail that electrons *lack* a simultaneous position and energy?', reflects this older view that electrons are in precisely defined states but

> that we (because of our crude techniques of investigation) cannot know what those states are; *we* cannot define them precisely although they *are* precisely defined. But this older view seeks what no physical theory can hope for – a knowledge of nature that transcends what our best hypotheses and experiments suggest.
>
> Quantum mechanics is a single, unified theory within which wave conceptions (or field or distribution or probabilistic conceptions) and particle conceptions (or point mass or singularity or granular conceptions) are equally fundamental in explaining and predicting the phenomena (Hanson [1972] p. 44).

The second stock objection to mind-brain causality seems to have worried Popper and Eccles somewhat more, as is evident in their tenth dialogue at the end of *The Self and Its Brain* (especially pp. 541 – 545). This is the claim that *if* interactionism were true, it would be in violation of the first law of thermodynamics, i.e. the law of energy conservation. But this objection, it seems to me, stems from a logical confusion. If the mind is a nonphysical entity, it cannot add to the total quantity of energy in the universe by acting on the brain, for it would have no physical energy to *transfer* (only matter having the potentiality of converting to energy). Similarly, if the brain acts upon the mind, this would not lead to a loss in the total quantity of energy in the universe, since an immaterial thing can hardly *absorb* physical energy the way material objects can[7].

Now Popper and Eccles, in the same dialogues that round out *The Self and Its Brain*, repeatedly suggest that we experience interaction between brain and mind all the time. But so far as I can see, what they mean by this is only that we experience gross mental-physical sequences of events, like wanting to raise my arm followed by its rising, or bumping into a desk followed by feeling pain in the thigh. When they try to make sense of that at the neuronal-perceptual level, they invariably stumble over the problem of giving explicit empirical content to the notion of observing the interaction between one's own brain states and one's *mind*. But this one could never do, if interactionism is true, for on its own terms the mind is not an observable entity, and certainly we cannot dissociate ourselves from our minds in order to observe them. We can in principle observe our brain states occurring, but not our perceiving of a stimulus as a result. We could all, in principle, observe brain states that lead to my finger moving voluntarily, but hardly the *wanting* to move it that caused those brain states.

But if this is correct, then it is clear that mind-brain interactionism, even if a true theory, is simply unverifiable. For no one other than oneself can have one's own perceptions or volitions, and if one cannot ever be in a position to observe one's mind and brain interacting, such interaction will never be detectable.

That a philosophical theory should be unverifiable, even in principle, is unsurprising. Yet it would be more satisfying if there were some way to get evidence for or against it, and even better if it could be seen to satisfy Popper's own criterion for demarcating scientific from non-scientific statements, i.e. that the theory be in principle *falsifiable*.

3. *Is interactionism falsifiable?* Several years ago it occurred to this author and to a sensory physiologist, R.W. Dykes, that while considerable progress was being made in mapping regions on the brain's surface that respond to stimulation of the various sense modalities, nothing had turned up that differentiated the tissue in those regions (other than contour and location on the neocortical surface) that correlated with the obvious dissimilarities between the modalities themselves. We studied three regions in particular: Brodmann's areas 3 (primary somatosensory cortex), 41 (primary auditory cortex), and 17 (primary visual cortex), in both cerebral hemispheres. After looking at the types of nerve cells, their arrangement in layers, and so on, we concluded that nothing in the known histology afforded even the beginning of an explanation of why feeling a touch, hearing a sonic boom, or seeing a glorious sunset are such different kinds of experiences[8]. The tissues involved, we concluded, are practically interchangeable. 'It is as if we discovered a single musical instrument that, depending on where it is situated in the orchestra, can produce equally well the sounds of a piano, a violin, and a bass drum'[9]. Nor, in the intervening five years, has this finding been overturned by new evidence: we still cannot point to anything in the brain that makes seeing subjectively different from touching [10]. Could futuristic experiments clear the matter up? As we wrote in 1978:

> Suppose we are able to overcome the present obstacle of central nerve regeneration in higher vertebrates and actually reconnect the auditory nerve with visual cortex and the optic nerve with auditory cortex in an adult dog or cat. If successful, there seem only two possible outcomes. Either the waking animal would have visual sensations upon auditory stimulation and *vice versa*, or this would make no difference to the animal's subjective experience and behavior. Either result would be baffling, but unequally so. We do not expect the latter, of course, because that would mean that visual cortex could take over auditory

> functions and *vice versa*, while the former result, in line with cortical stimulation and other phenomena, would mean that the population of neurons in area 17 can subserve only vision, and those in area 41 audition, no matter what the source of the impulses received. If so, what exactly predetermines their unique sensory function in each case? Or consider this more extreme speculation. Suppose in some future age of utopian neurosurgery we were able to *transpose* the tissue from area 17 to 41 and *vice versa*[11]. What would happen then? Would the animal experience flashes of light when we ring a bell in a darkened room and hear bells ringing when we flash lights in a soundproof room? That would imply that *something* in the tissue settles what sensory function it has, though with present-day knowledge we have no idea what that is. Worse still would be finding that the animal behaved normally after the transposition, since this would suggest, rather mystically, that just *being* in area 41 endows the tissue with hearing functions, as just being in area 17 gives it visual functions (Puccetti and Dykes [1978] p. 341).

Now the philosophical target of this argument was clearly all forms of mind-brain reductionism, since such theories uniformly express confidence that neurophysiology will eventually pre-empt psychology and in turn be pre-empted by physics and chemistry: the familiar physicalist's reverie. But there is also a more positive side to the matter. For upon reflection it seems clear that mind-brain interactionism is the only theory which predicts the failure to find dissimilarities in brain tissues to match the dissimilarities in sensory modalities subserved by those tissues. Interactionism does this, not just by making mind and brain distinct entities, but by giving mind its own functional role in information processing. In other words, above and beyond everything the brain does to decode afferent nerve impulses as information about its own bodily states and the extracorporeal environment, interactionism allows for a mind that *does further decoding of the neural input*, and it is this which accounts for striking sensory dissimilarities in spite of histological homogeneity in the primary receiving areas of the brain[12].

A corollary of this conclusion is most unwelcome to brain scientists: namely that, since on interactionism minds are unobservable, we shall never fully succeed in explaining the *qualia* of our sensory experiences, for the postulated further decoding of neural input is not available to intersubjective observation.

Yet, and rather strangely, this pessimistic finding is not so different from one put forward by some physicalists, which I should like to distinguish carefully.

The notion that what we perceive when we think we are perceiving an external world is really only states of our own brain is not confined to Popper and Eccles' speculations on the interplay between mind and cortical modules, already discussed. It is also found in neutral monists[13] and physicalists[14]. The paradigm is Russell's: the neurosurgeon when operating sees not the patient's brain but a portion of his own. Similarly Mountcastle, a neurophysiologist, says each of us lives within the 'prison' of his brain. Projecting from that prison are millions of fragile sensory fibres, in groups uniquely adapted to sample energetic states of the surrounding world. What reaches us via those fibres is all we directly know of that external world: all else is but logical inference. At higher levels of perception, in brain regions like the parietal lobe, each of us constructs his own very personal 'view from within'.

Elsewhere I have called this the 'I-am-a-Prisoner-of-my-Brain-Fallacy'[15]. What it overlooks is that our knowledge of the brain, including its filtering and integrative functions, is equally inferential from whatever experience we get of brains by studying them through microscopes, recording microelectrodes, EEG scans, staining techniques, and so on. So if we are going to be sceptical about the world beyond our brains, we have to be just as sceptical about what our brains are really like. Similarly, my brain does not provide me with an epistemologically unique vantage point. It is just as much an object in the external world to me as is your brain or anyone's. Maxwell claims that when we hear something we are hearing portions of our own brain, and the same with smelling. Now I can understand what it is to hear *with* my auditory cortex, but I cannot grasp what it would be like to *hear* my auditory cortex, unless one means just hearing the screech of amplified signals coming from a recording electrode implanted in my area 41.

But of course Maxwell did not mean that. He meant that somehow each of us is composed of a 'family of events', mental events, which are identical with some of the constituent events going on in our brain. For, following Russell, he finds no difficulty with the view that all of reality is composed of physical events organized into families, and that some of these are no less physical for also being mental, the way a man can be both a baker and a father.

I have already, in discussing Popper's *reductio* of the identity theory, agreed that what is wrong with this claim is that it makes the mental character of physical events in the brain causally redundant. But there is a further difficulty with the Russell-Maxwell version of the theory. For where is the 'I' who does the perceiving of his own brain? Maxwell, being a physicalist, wants to say that the 'I' is nothing apart from those constituent mental events that are also portions of his brain. Yet he needs the concept of the 'I' in order to state his theory in terms of it being *his* brain that he is is perceiving[16]. And that is precisely what he cannot supply on his theory. The primitiveness of the 'I' concept is what underlies our being able to understand Maxwell's talk of perceiving his own brain when he perceives the external world, but is not itself derivable from the notion of a family of mental-physical events in a brain, unless we already understand that the brain in question is indeed Maxwell's. By contrast, interactionism is congruent with this aspect of our conceptual structure, for it posits minds as possessing brains and in that way individuating them as having particular owners.

Quite apart from these theoretical advantages, however, it must not be overlooked that interactionism not only has the explanatory power to account for failure to find neuronal correlates of sensory disparities, but thereby opens itself to *empirical* refutation in the event such correlates are found[17] It is Karl Popper's distinction, along with John Eccles, to have mounted a thoroughgoing philosophical defence of the theory against its detractors. But the falsifiability of its claims elevates it to something even he did not expect: that is, an hypothesis of renewed scientific interest.

Notes

1. Popper and Eccles [1977] section 20, p. 74.
2. Popper wants to consider epiphenomenalism and psychoneural identity theory as more recent versions of classical *parallelism* for this reason. I think this is misleading terminology. In epiphenomenalism there is no causal order between mental events, and in psychoneural identity the mental and the physical can hardly be said to occur in parallel to each other.
3. Smart [1959].
4. Puccetti [1974].
5. Puccetti [1978].
6. Gazzaniga [1970].
7. For a discussion of how philosophers overstate the first law of thermodynamics in order to beg the question against interactionism, see Averill and Keating [1981].
8. Puccetti and Dykes [1978].
9. *Ibid*. p. 341.
10. Dykes, personal communication to the author, [1982].
11. The 'aura' preceding epileptic seizures, probably due to spread of abnormal patterns of neural activity, sometimes evokes sensations unrelated to external stimuli; interestingly, the subjective experience is still appropriate to the cortical area so recruited.
12. There is, of course, further decoding of the neural input within the brain beyond the primary receiving areas; but we claim that this too fails to account for sensory dissimilarities, whereas *mental* decoding could.
13. Russell [1948] and [1959].
14. Mountcastle [1975] and Maxwell [1976].
15. Puccetti [1978].

16. Strawson [1959].
17. At least within the problem context set out earlier. Should the empirical refutation transpire, interactionism would cease to have interest as a scientific hypothesis unless some other problem made it testable. It would, of course, retain its philosophical viability in view of the *reductio* it affords against physicalism.

4. SOCIAL FACTS AND PSYCHOLOGICAL FACTS

David Papineau

1. Karl Popper is probably best known amongst students of the social sciences for his 'methodological individualism'. Popper denies the existence of social wholes and urges that 'social phenomena, including collectives, should be analysed in terms of individuals and their actions and relations'[1].

In more recent years Popper has developed his views on the mind-body problem. Here he holds that mental processes are not reducible to (nor epiphenomena of, nor parallel to) physical processes; the crucial point is the way mental processes can involve entities from the (partially) autonomous third world.

It has occurred to some writers to wonder how Popper's views in these two areas mesh with each other. If reducibility goes for societies vis-à-vis individuals, why should it not go for minds vis-à-vis bodies as well? In particular the autonomy of the third world, on which the non-reducibility of mind to body hinges, seems little different from the notion of social fact defended by social holists[2].

My aim in this paper is to show that Popper is right to defend opposed positions in these two areas: while societies are indeed reducible, minds are not. I shall not, however, follow Popper's own route to these positions, but instead re-raise the issues in the light of some recent developments in the philosophy of mind and the theory of reduction. And it must be said that my defence of mind-body irreducibility is less full-bloodedly dualistic than Popper's. Still, there are a number of interesting intersections between Popper's original arguments and the line I develop, and I shall point them out as we reach them.

2. Let us begin with the question of mind-body reducibility. Once

Currie, G. and Musgrave, A. (eds.), Popper and the Human Sciences.
 ISBN 90-247-2998-X

upon a time there were perhaps materialists about the mind who thought that each *type* of mental event was identical to a certain *type* of physical event; that everybody who shared, say, a certain belief would thereby all satisfy some common physical description (no doubt of a type of brain state).

It is now generally agreed that such a type-type identity theory is grossly implausible. Whatever the relation between mind and brain, it is surely not true that in order for, say, President Mitterand and myself to share the belief that the sun has nine planets, we need to have an identical arrangement of neurones somewhere in our head. (And if gross implausibility is not enough, there is the possibility of prosthetic devices. It is likely that in the future we will be able to restore the mental capacity of people with damaged brains by artificially replacing bits of brain matter. The range of materials out of which such a device might be made is clearly quite open-ended.)

In place of 'type-type' identity theories we now have 'token-token' theories. One popular version (developed initially by Donald Davidson) has it that every *particular* mental event is identical to some *particular* physical event, even though there is no general match between all mental events of that kind and some physical description. On different occasions the same belief-type may be realized by quite heterogeneous types of physical state.

Such 'token-token' theories are normally considered to be versions of materialist monism. But it would not be entirely misleading to classify them as dualist. For even if they identify particular mental events with particular physical events, they allow (indeed insist) that mental *kinds* as such cannot be identified with anything at the physical level. And of course if mental kinds transcend the physical, then so too do instantiations of such kinds, mental states, mental facts, etc.

Still, an insistence on the non-reducibility of mental properties in itself commits token-token theorists only to a very minimal form of dualism. In the standard version the insistence on the autonomy of mental kinds is immediately tempered by various further theses downgrading the importance of such kinds. Most crucially, it is held that the laws sustaining causal relations between mental events and their various causes and effects are ones which apply in virtue of the physical descriptions of those events: it is their *physical*, not their mental, properties which give mental events their causal powers. And in addition (and independently) it is held that the mental 'supervenes' on the physical: no two situations can differ in some mental respect without differing in some physical respect. (It is important to realize that one can accept supervenience – each physical type fixes a certain mental type – without accepting the converse claim that each mental type implies a certain physical type. Note in particular that the disjunction of all the different physical types that fix a given mental type need not be finitely characterizable in physical terms.)

Most philosophers who count themselves as dualists about the mental are likely to demand something more than the minimal version offered in the Davidsonian scheme of things. (Popper's interactionism, for instance, certainly requires a less anaemic conception of the mental.) However, an analogous position might well seem attractive to those concerned to defend the autonomy of the *social*. In this area even those who feel sure that social facts escape individual definition, that somehow the features we pick out when we approach society in the large cannot be got at by building up from individual psychological description, generally feel uneasy about positing a realm of ontologically independent social facts. They would no doubt feel less uneasy if they could restrict their holist commitment to the thesis that social types (a competitive market, say) are not reducible to individual types (a specific set of attitudes on the part of the individuals involved), and simultaneously allow, firstly, that every particular social event is identical with some event composed of individuals (every particular competitive market is identical with certain individuals holding certain attitudes); secondly, that social events cause each other in virtue of generalizations involving their individual descriptions (competitive markets work as they do only in virtue of psychological principles about the individuals involved); and, thirdly, that two situations cannot differ socially without differing individually (if the nature of the market changes it must be because some of the individuals involved are different).

3. The problem I want to consider should now be clear. Maybe minimal, 'Davidsonian' dualism is too weak a position in the philosophy of mind, by comparison with some stronger form of anti-reductionism. But what about its analogue in the philosophy of social sciences? Here the weight of opinion (including Popper's) is *for* reduction. But why so? If the mental is entitled to more autonomy than Davidson offers, why should the social be entitled to less?

I shall proceed with this puzzle by considering a line of argument that a reductionist might mount against the Davidsonian stance. It will turn out that while this line of argument can be resisted in the case of the mind-brain relationship, it does succeed in discrediting the Davidsonian line on the social-individual case.

The basic idea is that the denial of type-type reducibility is in tension with our ability on occasion to predict certain mental events, and actions, on the basis of general assumptions relating them to other mental events described as such; and similarly with our ability to predict on occasion certain social events, on the basis of assumptions about their relation to other social events described as such. (In what follows I shall often simply take it as read that what I say for the mental goes for the social too.)

I realize there are those (Davidson, for instance[3]) who deny that we have any such abilities. They seem to me clearly mistaken. After all, the claim is not that we can *always* predict, merely that we *sometimes* can, as when it is completely obvious what someone will do, or inevitable that certain social consequences will ensue. There is of course more to be said on this point, but for the purposes of this paper I shall henceforth simply assume that we have at least *some* such predictive abilities.

Perhaps it is obvious why mental predictability is in tension with the denial of type-type reducibility. Nevertheless it will be helpful to spell it out. If certain mental antecedents, described as such, in general ensure a certain action, and if the rest of the Davidsonian apparatus is also held to be in operation, then the action is in a sense doubly determined on all occasions where it is so predictable, and there has therefore to be a certain co-ordination between the determinations.

Then suppose the mental antecedents, M, are realized on some particular occasion by physical conditions B, and that the action, A, is correspondingly then realized by physical conditions C. Then since A is supervenient on C, and C is determined by B (plus, no doubt, other physical factors present in the circumstances), A will be determined by (at least in the sense of inferable from) B, plus those other factors. But at the same time A is, by hypothesis, determined by (inferable from) M. So there is this constraint on the possible physical realizations of M on different occasions: they must all be such as (together with other factors) to determine (some C which determines) an action of type A. Now this is by no means an inconsistency. (Why *shouldn't* it always work out like that?). But given the denial of type-type reducibility, given the insistence that the possible realizations of M are physically quite heterogeneous, it is surely a massive coincidence. If on different occasions M is supposed to be realized in such physically disparate ways, then how come it always pans out that their physical consequences fix the same action?

It is obvious that just the same point can be made in the social case. If the different possible realizations of a given social event are supposed to be quite disparate at the level of their individual psychological descriptions, then how is that, in the case where there is some predictable consequence of the social event, they all conspire to give the same result?

4. One way of avoiding the tension between mental (social) predictability and the heterogeneity of physical (individual) realizations would be to drop the assumption of supervenience. If mental description (especially action description) no longer supervened on physical description, there wouldn't be any surprising co-ordination between the two realms in need of explanation.

But to drop supervenience now would be to opt for a quite unqualified dualism. We have already allowed that mental kinds enter into their own generalizations, by allowing predictability. This removes one possible argument for identifying particular mental events with particular physical events[4]. If we now accepted in addition that mental kinds are undetermined by physical description we would have no remaining basis whatsoever for denying the ontological distinctness of the mental realm.

So instead of following this path – which, again, might in the end be all right for the mental, but seems quite unattractive for the social – I shall consider a possible strategy for *reconciling* predictability with heterogeneity of realizations. The difficulty we faced was that there seemed no reason why all the different possible realizations at the physical level should all be such as to make the generalizations which allow predictability come out true. But suppose somebody maintained that it was simply part of our conception of the mental that this should be so. That is, might it not be that what we commit ourselves to by giving a certain mental description is simply the occurrence of some (any) realization which would make the relevant generalizations hold good?

The idea would then be that ascribing a mental state, while not laying claim to a specific kind of physical realization, does not leave the range of realizations completely open. By ascribing a given belief, say, we commit ourselves to a certain class of possible realizations. What class? Those realizations that will give rise to those actions and other mental events that we would have predicted on the basis of the original belief ascription.

This position in the philosophy of mind has got a name – functionalism. As such it is normally introduced by analogy with Ramsey's view of theories. According to Ramsey, a theory formulated with the help of theoretical terms is best viewed as a multiply existential claim with the variables ranging over unobservable properties: there exist certain properties which have such-and-such causal relations to each other and to observables (the 'such-and-such' being specified by the structure of the theory in question). So in the philosophy of mind the idea is to Ramsify our implicit psychological theory, and read it as saying that there exist in people certain physical states that have such-and-such causal relations to each other. And then to ascribe a given mental state to someone is to say that they are in some physical state with the relevant causal role.

So functionalism promises to accommodate the heterogeneity of mental realizations without giving up mental predictability. A given mental state, a given role in the cognitive causal structure, can in different organisms (or even in different people, or even in the same person at different times) be realized by different physical states, indeed by an open-ended range thereof. But the constraint on such realizations, that they all fill the appropriate causal role, does not leave it simply as a mystery that all these different possible physical realizations should conspire to the same consequences.

Again, it ought to be fairly clear how these ideas would apply in the case of the relation between the social and the individual. Indeed functionalism looks extremely plausible in this context. Suppose we start off with the idea that a sharp increase in the standard of living, or a sharp decrease therein, or a decline in the influence of the church, or any of a number of other changes, will all lead to a state of general *anomie*, and that this in turn leads to an increase in the suicide rate, to a rise in juvenile delinquency, to a rise in the number of destitute elderly, and to a number of other characteristic effects. Now we might want to allow that there might be any number of different realizations of this state at the level of individual psychology, that in different particular cases different particular sets of attitudes would be involved (feelings of personal inadequacy, resentment of different ethnic groups, straightforward existential anguish), having in mind only that, whatever the realization was, it would at least be one which resulted from (the realization of) an increase in the standard of living, or one of the alternative causes, and led in turn to an increase in the suicide rate, etc.

Again, we might have it that in a competitive market an increase in the cost of raw materials always leads to a re-assessment of production levels, and this in turn to a decrease in the supply of produced goods, but hold that the individual processes corresponding to that re-assessment are quite disparate on different occasions. Indeed this is something supported by empirical micro-research. Notoriously, when we actually look at what individual managers do, they do not conform to the picture painted by the ideal theory of the rational firm, and all adjust production until marginal revenue once more equals marginal cost. Instead they seem to adopt a variety of rules of thumb with no particular overall pattern. But nevertheless the large-scale effect remains constant: whatever the procedures used by the collection of individual managers involved in any particular market, they will at least be ones which will jointly produce a cut in production.

Again, to take one more example, we might have it that in industrial enterprises where the workers' representatives were regularly voted on by the workforce there were 'frictionless worker-management relationships' and therefore, say, less strikes involving disputes about conditions of work; but hold that the ways in which such 'frictionless relationships' might be realized are quite open-ended, subject only to the constraint that whatever they are they ensure less strikes of the kind mentioned.

5. For all its attractions, does functionalism explain the coincidence that I took to discredit the Davidsonian approach to the mind-body relationship? I have been speaking about the different possible realizations of a given mental or social state being 'constrained' to have a certain causal role. But do we not then need to ask *what* constrains the realizations all to have this causal role? And isn't this just the coincidence I was worried about in the first place?

The notion of constraint is deceptively ambiguous in this context. In the first place it can be understood in a definitional sense: the realizations of a given mental state are constrained to fill a certain causal role just in the sense that it would not *count* as a case of that mental state unless it had a realization with that causal role. This of course is the central tenet of functionalism, and as such I have no immediate wish to take issue with it.

But it is important to realise that if this is all that is meant by 'constraint', then functionalism quite fails to cope with the problem it was introduced to deal with, namely that of reconciling predictability with heterogeneity of realizations. For the definitional reading leaves it open that in order to identify someone's mental state one would need to attend its overall causal role, to check that it has the right overall structure of cause and effects. But if that were necessary then of course there would be no mental predicting, for we would need to know that certain effects occurred before we knew that we had a given mental state.

Since there is mental predicting this cannot be the whole story about our practice with mental concepts. I take it that what we actually do is identify mental states via some subset of their overall causal relations (maybe the presence of one of their possible causes, or of some symptomatic effect) and then go on from that to make predictions.

This does not necessarily mean that functionalism is wrong. Perhaps there is some sense in which our *concept* of a given mental state involves its total causal structure, so that if it ever turned out that the symptoms were present but the expected subsequent effects absent we would conclude we didn't have a case of that state after all.

But it does mean that if predictability is to be accounted for we need a stronger notion of constraints on realizations than the purely definitional one. That is, we need to know why it is that whenever we identify the symptoms of some mental state we will always have a realization that gives rise to the *further* expected effects. This is not just a matter of definition (since it is what allows predictability) and so, whichever way we cut it, there will still be the question as to why a heterogeneous set of physical states should always end up having the same effects.

Functionalists can of course say a given *belief* will always have the requisite effects, for that is built into their notion of that type of belief; but then the question is how certain symptoms always succeed in identifying that belief (identifying a state with certain effects), given once more that the realizations of that state are physically quite variable. Functionalism does not so much explain the coincidence as sweep it under the carpet.

This complaint might seem puzzling. Is not the functionalist doing with mental states just what Ramsey does with technical concepts in general? Yet the Ramsification of physical theories does not seem to generate accusations of inexplicable coincidences.

The reason is that the two cases are not entirely parallel. True, in the physical cases, just as in the mental, we identify the presence of various theoretical entities via some subset of their causal relations, and then go on to predict various effects. Predictability does hold in both cases. But heterogeneity of realizations does not. We don't in general think a given theoretical entity in physics is realized differently in different particular cases, and produces the results it does by quite different processes. We don't, for instance, think that the state of having a certain mass comes out differently on different particular occasions, and so gives rise to the accelerations it does because of different laws.

What, though, about cases where we *do* think we have different micro-realizations of a given macro-concept in physics? Temperature, say, can be treated as a theoretical concept à la Ramsey, and indeed a predictively important one, and yet we take it that the state of having a given temperature can be realized in an infinity of different ways at the level of particle movements. However it is crucial that despite this we do *not* here think that the different realizations are heterogeneous in the relevant sense. We do not think they have no common characteristic statable in the language of particle movements. It is precisely because we think that all realizations of a given temperature have the same mean kinetic energy that we consider the notion of temperature to be *reducible*. For it is because of this common characteristic that we can show by a *uniform* application of the laws of physics *why* the state of being at a certain temperature has the various effects it does.

It is just this kind of reducibility that I take to be denied by the thesis of the heterogeneity of realizations that is common coin between the Davidsonians and the functionalists. For someone who insists on this heterogeneity is not merely claiming that different realizations of a given mental state are physically distinguishable, as different realizations of a temperature are distinguishable. No type-type identity theorist would have wanted to deny *that*. Nor can they be saying that the different realizations have *nothing at all* in common – that is a nonsensically strong claim. What heterogeneity comes to is the precise claim that there is no finitely statable common physical characteristic *which in conjunction with the laws of physics implies that the state in question has the effects it does.* That is, heterogeneity amounts precisely to the claim that we can't, by looking at the nature of the physical realizations, explain why we always get the expected effects. But this claim – unless we move back towards a substantial dualism – is simply a statement of the coincidence I have been complaining of. If all the real causing is going on at the physical level, and the physical processes are that disparate, then how come we always get the same result? Functionalism might disguise the difficulty, but it doesn't remove it.

6. Not that functionalism is of no help at all. It does at least show us what still needs to be done if we are to reconcile predictability and heterogeneity of realizations.

What we need to do is to give some substantial account of what *does* constrain (in a non-definitional sense) all the disparate physical states which might realize a given mental state to coincide in respect of certain effects. In this section I shall suggest how this might be done. It will not be possible to give any more than an outline of this suggestion here, but it should suffice as a basis for my main thesis that what makes non-reducibility acceptable in the mental realm is absent in the social.

First let us note that at the physical level we are not simply random assemblies of molecules and neurones. On the contrary, we are rather carefully constructed by our genetic plans working in the context of our environments. Moreover, our genetic plans have been selected during the process of evolution to produce, in interaction with the relevant environment, characteristics which will have certain advantageous effects in that environment. In particular they have no doubt been selected to produce arrangements of molecules in the brain which are good at generating the appropriate behavioural responses in the appropriate circumstances.

However, for such arrangements of molecules to play such roles it obviously does not matter what precise molecular make-up they have, provided they establish the right structural connections between sensory inputs and behavioural outputs. And so it seems extremely plausible that our genetic plans will be such as to determine, not that there is a specific arrangement of molecules which will produce a given input-output link, but that there is *some* arrangement of molecules to do this. It is probable that the genetic plan selected during the evolutionary process itself proceeds by setting up its own selection process in turn, 'trying out' different arrangements of molecules, and 'opting' for whichever turn out to produce the 'right' results. And certainly one would expect that the exact arrangements of molecules we end up with will depend on such things as the order in which different abilities are acquired, on the language the individual speaks, etc. The genetic programme needs to get certain jobs done, but it will use whatever physical materials it has to hand.

Clearly if there is anything to the above story it will answer our difficulty. For the above story *explains* why a wide range of different physical states in different people all play the same causal role: it is, so to speak, the evolutionary purpose of those states to have that causal role.

Talk of purpose and design here is of course metaphorical, not to say controversial. But what ought not to be controversial is that however we describe it the evolutionary process produces organisms which work *as if* they had been designed by a purposive planner. So the fact that evolution manages to get different physical states to play the same causal role is no more mysterious than that neurosurgeons might find it appropriate to use different prosthetic devices in different circumstances to produce a given result. In neither case would the items in question have got there unless they played the roles they do – in the latter case because that is what the neurosurgeons wanted them to do, and in the former case because that is what our genetic programme has been designed to ensure.

Three further comments will help clarify the above argument:

(1) Firstly, on the connection between 'functionalism' as a position in the philosophy of mind, and the notion of 'functional explanation' as used in biology and elsewhere. When introducing functionalism in the former sense it is often necessary to start by disentangling it from the notion of functional explanation. A 'functional' property is a second-order property, the property of having some (any) first-order property satisfying certain restrictions. It is not necessarily a property which serves some purpose, and is thereby explained. There is no immediate reason for thinking the two notions have anything to do with each other.

There are of course those who think that functional explanation in biology is something to do with structures. But this is surely a mistake. Plenty of non-structural items have biological functions, and plenty of perfectly good structures aren't functionally explainable. There is of course more to be said on the logic of functional explanation, but for the purposes of this paper I shall take it that functional explanation in biology hinges on the already-mentioned circumstance that evolution 'selects' characteristics like a designer does (i.e. because of their effects)[5].

However it has now turned out that while functional explanation in general has nothing expecially to do with structures, the use of structural notions in psychology requires that the components of such structures have functional explanations. That is, we can usefully use a term for mental states in a second-order way, to identify that physical state, whatever it is, that has a certain causal role, only because such states have a function (namely, to play that causal role). For as we have seen, it is only because they are all eventual products of evolutionary design that the disparate collection of different physical states present in different people all end up working in roughly the same way.

(2) Secondly, a few remarks comparing my suggestion here with Daniel Dennett's views on the connection between psychological concepts and the logic of evolution. Dennett argues that we can apply the scheme of ideal rational decision theory to any organism, of a sufficient degree of complexity, that is the result of a process of natural selection. His thought is that *however such an organism's brain is structured* it will at least be the case that it will *somehow* be responsive to the relevant states of the environment, and that this will *somehow* inform its selection of the behaviour most likely to satisfy its needs[6].

No doubt there is something in this thought. But it is nothing much to do with the point I have been making. Whether or not assumptions of ideal rationality are in some sense the starting point for our understanding of ourselves and other organisms, I would argue *that for human beings at least* our (more or less implicit) psychological theory goes beyond rational decision theory to lots of detailed assumptions as to how and where we fall short of ideal rationality (assumptions about the limits on the computational complexity we can handle, about our tendencies to wishful thinking, etc.), and that it is accordingly predictively superior to mere ideal decision theory. Our psychological thinking doesn't stop with the thought that somehow we have *some* structure that approximates to ideal rationality; it adds specific hypotheses about the actual cognitive structure of human beings. But it doesn't go beyond that to hypotheses about the materials out of which such structures are made (not so much because we don't know, but because those structures will no doubt be differently realized in different humans). So my thesis is not that evolution ensures that however our brains are structured they will be roughly rational, but rather that it ensures that however our brains are physically realized they will have a certain specific structure.

(3) Popper also makes connections between the theory of natural selection and the nature of psychological states. For Popper it is the process of evolution by natural selection that leads to the emergence of problems and other World 3 objects. At a primitive level this happens even with non-human organisms, as when a pair of birds are faced with the problem of finding a good nesting place; it comes to full fruition with the development of language and the further problems, theories

and logical relations this generates. And it is the partial autonomy of the third world objects that thus emerge that for Popper implies the irreducibility of psychological to physical states[7].

Again, this is not the use I wish to make of the theory of evolution. As mentioned earlier, Popper argues for a much stronger form of irreducibility than that I am concerned to defend, and it is not my purpose here to pursue his line of argument. Instead I shall now finally turn to the task of showing why Popper is right to deny even the more limited form of irreducibility of social entities.

7. The moral should now be pretty obvious. We had the social anti-reductionist appealing to functionalism along with the mental anti-reductionist, as a way of reconciling heterogeneity of realizations with predictability. But it turned out that functionalism failed by itself to resolve the tension between heterogeneity and predictability. We were still faced with the puzzling circumstance that the quite different physical states (or in the social case, individual states) that we identify on different occasions as a given mental state (social state) should all coincide in the results they produce.

One way of resolving the tension would of course have been to abandon heterogeneity, to allow that there is something physically (individually) in common between all the different realizations, which together with general truths of physics (psychology) explains why they all have the same effects. But we were able to resist this reductionism for mental states by giving an alternative explanation for the coincidence functionalism left us with: evolution by natural selection ensures that certain causal roles get played, even though different physical states might get picked for the part.

It is not at all clear that anything analogous offers itself in the social case. Different societies aren't given genetic programmes by some process of natural selection to ensure that despite their heterogeneity at the level of individual psychology they will be causally similar at the macro-level. And so if we do continue to believe in some such causal similarities, if we do accept that we can sometimes predict on the basis of information at the macro-level, then there seems little alternative to reductionism, to the admission that our procedures for applying such terms as enter into macroscopic social generalizations will guarantee uniform realizations at the psychological level, of a kind to which the application of psychological generalizations will predict uniform results.

Of course we needn't *know* the reduction in order to have knowledge of some macro-generalization (though we often will[8]). All this talk of reduction has been ontological rather than epistemological. But note that the argument does imply that the acceptance of a macro-generalization commits one to there *being* some such reduction.

Perhaps all this is a bit quick. Maybe societies aren't given genetic programmes by a process of biological natural selection, but could there not be some other kind of selective process to ensure that disparate individual states realizing a given macro-description all play the same causal role? Thus in our economic example it might be argued that the exigencies of the competitive market ensure that however individual managers behave, and there might be a wide range of such behaviours, they will at least act in ways which ensure that their firms make money (otherwise the firms would long since have disappeared from the scene), and thus that we can be assured that while the psychology of individual businessmen is varied and curious it will still be such as to ensure various joint effects.

And of course there is the much more obvious kind of case, illustrated in the industrial relations example, where it is as a result of conscious choice and selection that the fillers of certain roles, despite differences of psychological make-up, are always ones who produce certain effects – the voters don't mind how the representatives go about getting on with the management, as long as they somehow succeed in communicating.

I do not intend here to adjudicate how far predictability can be reconciled with heterogeneity in the social realm by appeal to such alternative selection processes. I myself think that the two kinds of cases just mentioned are rather special, and that as a rule the acceptance of social generalizations commits one to the existence of a (perhaps unknown) reduction. Certainly in the anomie example sketched earlier there seems no room for any appeal to any selection process, and so if we think 'anomie' signifies a social state with certain standard effects we shall need to accept it has a standard kind of individual realization as well.

One last observation. There is in effect already a significant amount of literature on precisely this question, of how far social phenomena result from selection processes. For if it is accepted that functional explanation in biology hinges on allusions to evolution by natural selection, as I suggested earlier, then the acceptability of functional explanations in the social science requires the existence of analogous selection mechanisms. And a number of writers have discussed functional explanation in the social sciences in precisely these terms, with the general consensus that, the conscious design case aside, the possibility of such functional explanations is fairly limited[9]. My argument has been that if they are right then the possibility of avoiding social reductionism is equally limited.

Notes

1. Popper [1963] p. 341.
2. See for instance, R.J. Ackermann [1976] chapter 9.
3. Though see Essay 14 in D. Davidson [1980].
4. For the argument, see Essay 11 in D. Davidson [1980].
5. Cf. L. Wright [1976].
6. D. Dennett [1978], *passim*. See also D. Dennett [1981].
7. See in particular Ch. 6, Popper [1972].
8. Cf. p. 117, D. Papineau [1978].
9. See for instance R.P. Dore [1961].

5. METHODOLOGICAL INDIVIDUALISM: AN INCONGRUITY IN POPPER'S PHILOSOPHY

A. F. Chalmers

1. *Introductory remarks.* Karl Popper recommends that when theorizing about the nature of societies and social change we should adopt a standpoint that he refers to as 'methodological individualism'. Although methodological individualist social theories are individualist theories in some sense, their individualist character should be distinguished from a brand of individualism that Popper, following Husserl, labels 'psychologist.' Psychologism, a standpoint that Popper emphatically rejects, is 'the plausible doctrine that all laws of social life must be ultimately reducible to the psychological laws of "human nature" '[1], a standpoint which J.S. Mill adopted when he claimed that 'all phenomena of society are phenomena of human nature' so that 'the Laws of the phenomena of society are, and can be, nothing but the laws of the actions and passions of human beings'[2]. In taking his stand against psychologism Popper advocates an 'autonomous sociology' distinct from psychology.

In this article I will argue that there are two strands to, or aspects of, Popper's anti-psychologistic methodological individualism. One of them, which I will refer to by Popper's phrase 'autonomous sociology'[3] is one for which Popper offers strong arguments and is one which I basically accept as having a domain of application. I shall argue that it is inappropriate and misleading to refer to autonomous sociology as individualist. The second strand to, or aspect of, Popper's methodological individualism is appropriately referred to as individualist. I will call it 'reductive individualism'. Reductive individualism is a position I wish to reject and, moreover, it is a view that clashes with much of the spirit, and sometimes the letter, of Popper's own writings.

2. *Against psychologism.* Before presenting my elaboration and cri-

Currie, G. and Musgrave, A. (eds.), Popper and the Human Sciences.
 ISBN 90-247-2998-X

tique of Popper's methodological individualism, I will summarize and attempt to clarify his case against psychologism, a case that I accept.

One problem faced by psychologism is the difficulty of establishing precisely what the 'laws of human nature' are and of distinguishing psychological characteristics that are due to human nature from those that are a product of a specific society and that are induced in humans by virtue of their actions and experiences within that society. Popper points out that there is evidence that even some apparently near-universal 'instincts' such as the fear of snakes are socially induced rather than being inherent in humans by virtue of their nature[4]. Psychologism requires a theory of human nature that we do not possess and which, for methodological reasons, we may never possess.

A second, more serious, objection to psychologism stems from the fact that, in order to be carried through consistently, a psychologistic theory of society is committed to tracing the generation of a particular society back to the origins of human society. The reason for this is as follows: individuals are born into a society that they do not choose or create. Many of their psychological characteristics will be moulded by their actions and experiences in that society. Further, an action of an individual will not be fully understood without reference to the situation in which the individual acts. Reference to psychological desires or motives will not be sufficient. A defender of psychologism must respond to these elementary observations by insisting that the social context in which an individual acts and by which he or she is influenced is man-made and is itself to be explained by a psychologistic reduction. However, this initiates a train of reasoning that takes us on a regress back to the origins of society, to the very first society which arose, so the defender of psychologism must hold, by virtue of the properties of human nature alone.

From the standpoint of psychologism, social theory is certainly practically inaccessible to us because of the impossibility of carrying out the necessary reduction. There are also powerful reasons for doubting the assumption that all social developments from the origins of society up to the present are to be explained by reference to the psychology of humans. These reasons include the effect of economic and material conditions on social development and the fact that many aspects of our environment 'are, as a rule, *the indirect, the unintended and often the unwanted by products*' of our actions[5]. We can also add Popper's observation that there are reasons to believe that our human ancestors were social prior to being human, so that the very first step in the psychologistic reconstruction is suspect[6].

Some aspects of Popper's antipsychologistic view of social development show a parallel with his views on knowledge and its growth. According to Popper,

> ... the *tabula rasa* theory is absurd: at every stage of the evolution of life and of the development of an organism, we have to assume the existence of some knowledge in the form of dispositions and expectations.
>
> Accordingly, *the growth of all knowledge consists in the modification of previous knowledge* – either its alteration or its large-scale rejection. Knowledge never begins from nothing, but always from some background knowledge – knowledge which at the moment is taken for granted – together with some difficulties, some problems (Popper [1979] p. 71, italics in original).

An individual is born into a world in which there already exists much knowledge, which has various features whether he or she realises or likes it or not and which he or she can attempt to contribute to and modify by available means. Similarly, an individual is born into a society that has various features, whether he or she realises or likes it or not, and which he or she can attempt to contribute to or modify by available means.

3. *Autonomous sociology*. In this section I will elaborate on that part of Popper's defence of an autonomous sociology that is compatible with his anti-psychologism.

In this attack on psychologism Popper insists that individuals within a society are confronted by a social situation, a situation that cannot usefully be analysed in terms of the psychology of the humans that are involved in or have given rise to that situation. In the light of this it would seem that the role of sociology is to theorise about the social situations themselves, and hopefully to give guidelines concerning how those situations can and cannot be changed. There are a number of passages in Popper's writings which support precisely that view.

A striking example follows immediately upon Popper's criticism of psychologism discussed in the previous section. Popper insists that psychologism 'fails to understand the main task of the explanatory social sciences.' He then proceeds to explain that task as follows:

> (The task) is ... the discovery and explanation of the less obvious dependences within the social sphere. It is the discovery of the difficulties which stand in the way of social action – the study, as it were, of the unwieldiness, the resilience or the brittleness of the social

> stuff, of its resistance to our attempts to mould it and to work with it (Popper [1966] volume 2, p. 94).

and a page later he writes,

> Social life is not only a trial of strength between opposing groups: it is action within a more or less resilient or brittle framework of institutions and traditions, and it creates – apart from any conscious counter-action – many unforeseen reactions in this framework, some of them perhaps even unforeseeable (*ibid.* p. 95).

Elsewhere, in a similar vein, Popper talks of the need for '*situational analysis* which provides us with the models (rough and ready to be sure) of typical social situations'[7].

The emphasis, then, is on the analysis of social situations, of the 'social stuff' as opposed to an analysis of individuals. There are a number of themes in Popper's writing on the social sciences entirely in keeping with that emphasis. One of them is Popper's emphasis on the unintended consequences that almost invariably result from human action. '(Only) a minority of social institutions are consciously designed while the vast majority have just "grown" as the undesigned results of human actions'[8]. Consequently, writes Popper elsewhere, on the same topic,

> It is the task of social theory to explain how the unintended consequences of our intentions and actions arise, and what kind of consequences arise if people do this that or the other in a certain social situation. And it is, especially, the task of the social sciences to analyse in this way the existence and the functioning of *institutions* (such as police forces or insurance companies or schools or governments) and of social *collectives* (such as states or nations or classes or other social groups) (Popper [1969] p. 125).

Allied with this position on unintended consequences is Popper's rejection of the conspiracy theory of society, according to which social phenomena can be explained by reference to the individuals or groups of individuals who have conspired to bring them about[9].

A useful way to illustrate Popper's notion of an autonomous sociology is by reference to Marx's analysis of what Popper refers to as 'unrestrained capitalism', many aspects of which Popper claims to accept and admire. To quote Popper himself,

> (Marx's) attempt to use what may be called 'the logic of the class situation' to explain the workings of the institutions of the industrial system seems to me admirable, in spite of certain exaggerations and the neglect of some important aspects of the situation; admirable, at least, as a so-

> ciological analysis of that stage of the industrial system which Marx has mainly in mind: the system of 'unrestrained capitalism' ... of one hundred years ago (Popper [1966] volume 2, p. 117).

Marx's analysis of the relationship between the classes within an unrestrained capitalist society illustrates the anti-psychologistic aspect of his theory. The ruling class, the capitalists, own the means of production. They produce and sell on an open market in competition with other capitalists. The capitalist must sell at the lowest possible price. If he does not he will be forced off of the market by those of his competitors that do. Consequently, he is forced, by the nature of the situation, to expand and improve the means of production and to maintain the wage of the workers as low as possible, that is, at subsistence level. For their part, the ruled class, the workers, since they have no direct access to the means of producing their livelihood, are forced to work for the capitalist for whatever wages they can get. This model, when suitably elaborated, can explain, for example, why, in mid-eighteenth century European society, vast improvements in the industrial productive forces did nothing to alleviate the large scale suffering and starvation amongst the working class[10].

In his philosophy of the natural sciences Popper insists that laws can best be understood as prohibitions, asserting that such and such a thing cannot happen, as exemplified by the proverb, 'you cannot carry water in a sieve'[11]. Given the Marxist account of the capitalist mode of production it is possible to deduce some laws of that kind applicable to unrestrained capitalist societies. For example, it is possible to show that, in an unrestrained capitalist society with a surplus population, 'wages cannot rise above starvation wages'[12]. It is also possible to show that the plight of the workers cannot be solved by a moral campaign urging capitalists to be more generous to them. Capitalists succumbing to the campaign will go broke at the expense of those that do not. This last example brings out the point that the theory in no way depends on the attitudes or other psychological attributes of individuals. Given an unrestrained capitalist system, the capitalist must compete or go broke at the expense of his competitors and the workers must work for what wages they can get or starve.

At this point it is pertinent to raise the question of the appropriateness of referring to the autonomous sociology supported and exemplified in Popper's writings as individualist, since much of what has been said in this section suggests that it is inappropriate.

There is one clear sense in which, according to Popper's view, any society is individualist. No society would function, or even exist, if it were not for the presence and actions of the individuals that make it up. Further, if a society is to change, that change must in some way be brought about through the actions of individuals. For example, if the unrestrained capitalist society discussed above is to exist and function it is necessary that some, though not all, of the workers be prepared to work rather than starve and it is necessary that some, though not all, aspiring capitalists do what is necessary to remain competitive. In taking this stand Popper is opposing holist or collectivist theories of society according to which society is animated by some driving force other than individual human beings, such as a collective mind (Rousseau) or an absolute spirit (Hegel). Popper is opposed to the view that social change is subject to a law of development uncontrollable by humans[13]. However, that agreement with Popper here does not commit one to what can usefully be labelled an individualist theory, can be brought out with an analogy. It is undoubtedly the case that a bicycle needs an individual to 'animate' it. In theorizing about, or explaining the workings of, a bicycle we will focus on the mechanics of the bicycle, the physics of centrifugal forces, gear wheels and so on. Our analysis will take for granted that the bicycle will be pedalled and steered by some human. What is appropriate for the rider to do will be determined by the design and workings of the bicycle. Analogously, an autonomous sociological explanation of some aspect of society, such as the relationship between the classes in an unrestrained capitalist society in our example sketched above, will presuppose that the system will be activated by humans acting in a way more or less appropriate to the situation. A theory of society that attributes to humans a function they cannot in fact perform, as is the case in the characterization of some Utopian societies, will be a bad theory for that reason. I suggest that to label a social theory 'individualist' because it presupposes that individuals are necessary to animate it is as appropriate as calling a theory of the operation of bicycles individualist because it makes a similar assumption[14].

4. *Reductive individualism.* If this were all that there is to be said about individualism in Popper's writings on the social sciences, then my paper could end here with the remark that Popper has outlined some aspects of an anti-psychologist, non-holist approach to social theory which has a legitimate domain of application and which is exemplified in much of Marx's work but which Popper misleadingly

refers to as 'methodological individualism'. That the matter cannot be left here is due to the fact that there is evidence in Popper's writings of a much stronger kind of individualism which I will argue is out of keeping with his autonomous sociology and which should be rejected.

The individualist strain in Popper's writings, which I will refer to as reductive individualism, involves the assumption that all social phenomena must be reduced to properties of and relations between individuals. This point is made quite explicitly by Popper when he writes that 'the task of social theory is to construct and to analyse our sociological models carefully in descriptive or nominalist terms, that is to say, *in terms of individuals*, of their attitudes, expectations, relations, etc. – a postulate which may be called "methodological individualism" '[15]. Popper insists that the valid aspect of psychologism lies in the fact that 'it rightly insists that the "behaviour" and the "actions" of collectives, such as states or social groups, must be reduced to the behaviour and to the actions of human individuals'[16]. Thus, according to Popper, collectives such as 'the army' are abstract concepts and 'we are liable to feel that we see [them], either within or behind the changing observable events, as a kind of permanent ghost or essence' whereas what is concrete is 'the many who are killed; or the men and women in uniform, etc.'[17]. Popper refers to the 'popular view that social entities such as institutions or associations are concrete natural entities such as crowds of men, rather than abstract models constructed to interpret certain selected abstract relations between individuals'[18].

If we take Popper's remarks to the effect that we must reduce social phenomena to the 'attitudes, expectations, relations etc.' of individuals at face value then it is difficult to see how this position can be free of psychologistic elements, for it is surely the case that analysis of the attitudes and expectations of individuals must involve psychological considerations. However, Popper insists that his methodological individualism differs from psychologism[19]. It would seem that the way in which Popper sees his reductive individualism differing from psychologism is that the former relies, not on the psychological attributes of individuals but on their *rationality*[20]. I will postpone a discussion of this issue and of Popper's use of his 'rationality principle' until the next section. In the remainder of this section I will explore some arguments against reductive individualism *most of which can be gleaned from Popper's own writings.*

It is difficult to see how reductive individualism can escape what I see as the central argument against psychologism. Individuals are confronted by a social situation and their behaviour and actions and the consequences of their behaviour and actions cannot be understood without reference to that social situation. Consequently, *any kind* of individualist reduction would seem to demand that the social situation that an individual confronts be reduced to the behaviour and actions of humans. We are forced into the regress back to the origins of society that Popper shows psychologism is forced into, with the minor difference that Popper has the option of explaining that origin in terms of the rationality rather than the psychology of humans, a step that does little to offset the force of the objection.

It would seem that if we are to characterize and analyze the social situations confronted by individuals in a way that avoids the regress, then we must attribute to the entities involved in a characterization of the situation, such as the state, the army and the means and relations of production, a status that cannot be 'reduced to the behaviour and to the actions of human individuals'. Popper virtually admits as much when, in those passages I used to illustrate his support of an autonomous sociology, he writes of the task of social theory to study 'the unwieldiness and resilience or brittleness of the social stuff', of social life as 'actions within a more or less resilient or brittle framework of institutions and tradition', of the growth of institutions as the undesigned results of human actions, and of the need for the social sciences to analyze 'the existence and functioning of institutions ... and of social collectives'. What is more, those aspects of Marx's analysis of unrestrained capitalism that Popper admires and accepts do not conform to the demand for an individualist reduction.

Popper's reluctance, when defending his reductive individualism, to ascribe significant ontological status to social entities is curiously at variance with the permissive attitude towards ontological matters that he shows in his stance on the status of human knowledge. Popper insists that scientific theories, problem situations and arguments and the content of libraries, books and journals are ontologically distinct from physical objects and from the subjective thoughts or beliefs of individuals[21]. They inhabit a '*largely autonomous* third world of objective knowledge'[22]. Objective knowledge is a structure that confronts individuals and which they can work on and add to 'as masons work on a cathedral'[23]. One of Popper's arguments for the autonomy of the third world of objective knowledge is the unintended consequences of attempts to modify and add to it, for 'the moment we have produced these theories, they create new, unintended and unexpected problems, autonomous problems, problems to be discovered'[24]. The third world is autonomous and is not susceptible to a reduction to the beliefs or mental states of individuals, in spite of the fact that 'it is to be admitted that the third world, the world of objective knowledge ... is man-made'[25].

It would seem that Popper's arguments for the autonomy of the third world apply with equal force to the social world, especially in the light of Popper's emphasis on the situation that individuals confront and the unintended consequences of their actions in those situations. Indeed, insofar as human knowledge, and such things as written constitutions and laws, form part of the social situation, Popper's views on the autonomy of the third world are inconsistent with his reductive individualism and with his explicit refusal, for instance, to attribute an ontological status to the army over and above the individual soldiers that are its members. It would seem that Popper could well have written a sequel to his article 'Epistemology without a knowing subject' with a title something like 'Social change without a designing subject'. Such a move would be in keeping with his autonomous sociology but would, of course, clash with his reductive individualism.

Another incongruity, if not inconsistency, in Popper's writings consists in his insistence on an individualist reduction within sociology, on the one hand, and his reasons for rejecting an analogous insistence that mental states be reduced to physical states on the other. Popper's case for rejecting insistence on this type of reduction draws on the example of the reduction of chemistry to physics. Popper observes that, had it been insisted, from the outset, for philosophical reasons, that chemistry must be reduced to physics, and that talk of chemical states is illegitimate unless they can be reduced to physical states, then advances in chemistry would have been ruled out from the outset. If chemistry were to be completely reduced to physics, as seems likely, then this of course would constitute a major achievement, but a precondition for that achievement is the existence of a developed chemistry in the first place. On the other hand, if the reduction were to prove impossible, then the absence of a developed chemistry would clearly be a loss. Either way, it is advantageous to develop chemistry in its own right and not to demand a reduction from the outset[26]. The moral Popper draws from this with respect to the mind-body problem is that 'we must be pluralists, at least to start with'[27] whilst, if we seek reduction 'we must in each case try to learn as much as possible about the field which we hope to reduce ... and in some cases, we may even possess arguments to show why the field cannot be reduced'[28]. The policy Popper advocates here tells against his support of reductive individualism and, indeed, we have discerned evidence in Popper's own writings that we 'possess arguments to show why (social theory) cannot be reduced (to talk of individuals)'.

As mentioned in the previous section, Popper is concerned, in his social theories, to combat holism, and any theory that implies that society is animated by some force other than the actions of humans, such as a group mind or spirit. My analogy between the workings of society and the workings of a bicycle brings out the point that reductive individualism is much stronger a position than is needed to combat holism[29]. Bicycles need human beings to activate them, but their behaviour cannot be *reduced to* the behaviour and actions of humans. The same point can be made with reference to Popper's conception of human knowledge. Human knowledge is generated by humans and can grow and develop only through the intervention of humans. Nevertheless, that knowledge has some kind of autonomy that renders an individualist reduction impossible. A similar view can be held with respect to society, in conformity with Popper's autonomous sociology but conflicting with his reductive individualism.

5. *Situational analysis and the rationality principle.* On occasions, Popper has talked of his approach to social theory in terms of 'situational analysis' and the 'rationality principle'[30]. In this section I will argue that part of what Popper is getting at can be reformulated in a way that is in keeping with his autonomous sociology, whilst the other part, especially Popper's talk of the rationality principle as applied to individuals, is allied to his reductive individualism and is more dubious.

Popper illustrates his use of situational analysis and the rationality principle in social theory by drawing an analogy with our explanation of the solar system in physics. The description of the physical situation involving planets moving in approximately elliptical orbits around the sun is taken as analogous to the characterization of some social situations. Having drawn this analogy, Popper poses the question of the ways in which these models are animated. In the case of the solar system Newton's laws of motion are deemed to be responsible for keeping the system running. In the case of a social situation, 'we need, in order to "animate" it, no more than the assumption that the various persons or agents involved act *adequately, or appropriately*; that is to say, in accordance with the situation'[31]. It is this assumption, that individuals act appropriately to the situation in which they find themselves, that Popper calls the 'rationality principle'.

By introducing his universal rationality principle in this way, Popper is able to construe sociology as having the same formal, hypothetical-deductive structure as the one he attributes to physics, in keeping with his views on the unity of method[32]. However, as Noretta Koertge has pointed out[33], there are fundamental disanalogies between Popper's conception of physics and sociology practised in terms of situational analysis and the rationality principle. From the point of view of Popper's analysis of the method of the physical sciences, Newton's theory has a high standing as a scientific theory because of its high content and hence its high degree of testability or falsifiability. What is more, Popper's methodology demands that we should ruthlessly test Newton's theory in an attempt to discover its shortcomings. If we are successful we should attempt to replace Newton's theory by a bolder, more adequate conjecture designed to solve the problems unearthed by our criticism. The treatment that Popper proposes for his rationality principle, which he introduces as the analogue of Newton's laws in social theory, is markedly different. In the first place, he admits that the principle is false[34]. Individuals often act in a way that is inappropriate to the situation in which they find themselves. However, Popper does not conclude from this that we replace the rationality principle by a bolder, more adequate conjecture. He recommends that the principle should be protected from falsification by the methodological device of always blaming our situational analysis rather than the rationality principle for the breakdown of our social theories. This must sound very much like taking a dogmatic attitude towards the applicability of the rationality principle to a serious Popperian critical rationalist. Further, Popper's recommendation that 'we should pack or cram our whole theoretical effort, our whole explanatory theory, into an analysis of the situation' rather than focus on the rationality or otherwise of the agents involved[35] whilst in keeping with his autonomous sociology, is the converse of what one would expect of one who purports to support an *individualist* theory of society.

Popper's emphasis on situational analysis, then, is quite in keeping with the autonomous sociology I have extracted from his writings, whilst from that standpoint his rationality principle poses problems. I suggest that the function of the rationality principle within Popper's sociology is to constitute a denial of holism. Societies are animated by humans and not by a group mind or spirit. The denial of holism can be achieved by replacing Popper's rationality principle by a much weaker one. We need simply claim that, if a society or social situation is to function or change, then some humans must do what is necessary for it to function or change. I will refer to this weaker alternative to Popper's rationality principle simply as the animating principle. Apart from a denial of holism, this principle has little precise content until it is filled out by some detailed specification of a social situation. Once such a specification has been given in a particular case it will be possible to spell out the tasks that need to be carried out by humans in that case. For example, the Marxist characterization of unrestrained capitalist society, discussed above, indicates the tasks that need to be carried out by workers and capitalists. It is important to note here that the analysis does not involve any specification of the actions of identified individuals. It requires that a number of capitalists compete with each other in appropriate (specifiable) ways and that a number of workers work and reproduce to keep the wheels of production turning. The theory is not threatened by the fact that some capitalists will act inappropriately and become bankrupt whilst some non-capitalists will starve rather than work for subsistence wages. Nor is it necessary to specify which individuals act in a way that is appropriate to their situation and those that will not, nor to say anything about their motives or other psychological states. Any specification of a social situation, taken in conjunction with the animating principle specified above, will have the consequence that some, unspecified, humans act in ways appropriate to the situation. Instantiations of this assumption will often be uncontroversial. For example, the analysis of unrestrained capitalism involves the assumption that, faced with the option of working for subsistence wages or starving, a significant number of workers will do the former.

Replacement of Popper's rationality principle, that individuals always act in a way appropriate to their situation, by the much weaker animation principle, that if a social situation is to function then some humans must do what is necessary to make it function, is quite compatible with the development of a non-holist autonomous sociology that places an emphasis on social situations in their own right. The move has the advantage that it avoids some of the problems concerning the interpretation and status of Popper's rationality principle and it also blocks the way to understanding and employing the rationality principle in a way that is out of keeping with autonomous sociology and more in keeping with reductive individualism, a tendency that is present in the writings of Popper and some of his followers.

If the reductive individualist strand in Popper's writings is taken seriously then it is necessary to pursue the question of the precise status and content of the rationality principle and the way in which it can be employed to explain the actions of individuals. We have already noted the difficulty posed by the fact that the principle is false. Popper responds to this observation with the claim that it is 'as a rule, sufficiently near the truth' although he offers no support for this latter assertion. Koertge has suggested that the principle is inadequate for situations involving irrational fears, desires and emotions[36]. An ambiguity in Popper's formulation of his rationality principle concerns the question of whether individuals are claimed to act in a way that is appropriate to the situation as it is or as they see it. At one place Popper indicates that the latter corresponds to his intentions. He asserts that the rationality principle 'assumes no more than the adequacy of our actions to our problem situations *as we see them*'[37]. Popper talks in these terms of explaining the actions of specific war leaders and even of madmen[38]. However, once we have reached the stage of subsuming the actions of madmen under the rationality principle it is difficult to see how the principle can play a fundamental role in social theory. We have come a long way from autonomous sociology. In this respect, it is interesting to contrast Koertge's response to the problems posed by Popper's rationality principle with my own. Koertge identifies the problem of explaining human actions in terms of the rationality principle as the core of 'Popper's research programme in the social sciences'[39], and recommends that the principle be *strengthened* so that it becomes more falsifiable. My own recommendation is that Popper's autonomous sociology can be improved by replacing the rationality principle by a *weaker* animating principle. As for explaining the specific actions of individuals, some version of the rationality principle might have some application here, but that has no relevance for the core of Popper's research programme in the social sciences if we take that core to be his autonomous sociology rather than his highly problematic reductive individualism.

6. *A disclaimer.* In this article I have considered the problem of methodology within the social sciences in Popper's own terms and I have located what I have argued to be confusions and incongruities in his position. However, I have reservations about the nature of the

debate. The reservations concern the extent to which one can expect to be able to arrive at general answers to questions concerning what does and does not constitute good social theory at the level of philosophical or methodological analysis. Philosophy does not have the resources that make it possible for it to legislate on what standards social theories of the future must conform to. Rather, I suggest that we will learn what can be achieved in the domain of social theory by developing successful ones. If, in this article, I have shown myself to be more sympathetic to autonomous sociology than to reductive individualism that is because the former is exemplified by a notable achievement in social theory, namely, the analysis of unrestrained capitalist society achieved by Marx and acclaimed by Popper.

My position with respect to the relationship between the methodology of social theory and social theory itself can be illustrated by a comparison with the situation with respect to the physical sciences. Galileo and Newton demonstrated how general theories formulated in mathematical terms can be applied to the physical world. Prior to their achievement there were philosophical arguments, many of them formulated by Aristotle, to the effect that mathematical descriptions are not applicable to the, in many respects chaotic and unordered, physical world[40]. The Galilean solution to the problem involved the replacement of observation by controlled experiments designed to isolate systems capable of exhibiting, or yielding evidence bearing on, the mathematical order described by the new mathematical physics. Philosophy did not, and could not, predict the success of the efforts of Galileo and Newton. This is in keeping with Lakatos's observation concerning the disparity between Newton's scientific achievements and his theory of that achievement. 'He and his brilliant followers established, *in practice*, the basic features of scientific methodology'[41] It has taken several centuries for anything like an adequate account of that methodology to emerge. Future breakthroughs in social theory may force autonomous sociologists, individualists and holists alike to change their views on what kinds of theories of society are possible, just as the innovations of Galileo and Newton forced Aristotelians and, eventually, classical empiricists, to change their views.

Notes

1. Popper [1966] volume 2, p. 89.
2. As cited by Popper (*ibid* p. 91).
3. I prefer 'social theory' to 'sociology' in view of the dubious character of much of contemporary theory that is referred to by the latter name.
4. *Ibid.* pp. 89-90.
5. *Ibid.* p. 93, italics in original.
6. *Ibid.* p. 93.
7. Popper [1967] p. 143, italics in original.
8. Popper [1961] p. 65. The whole sentence is italicised in the original. See also Popper [1966] volume 2, p. 93.
9. See Popper [1966] volume 2, pp. 94 – 5 and Popper [1969] pp. 123 – 5.
10. Popper fails to acknowledge the extent to which Marx realized that his basic theory was but the bare bones of a model that needed to be modified and augumented to take account of the complexities of any actual capitalist society. For documentation of this point see Suchting [1972].
11. See Popper [1961] p. 61 and Popper [1972] section 15.
12. Popper [1966] volume 2, p. 178.
13. See Popper [1966] volume 2, p. 91 and Popper [1961] p. 149.
14. I would not wish my analogy to be pushed too far. There are clearly important differences between societies and bicycles.
15. Popper [1961] p. 136, italics in original.
16. Popper [1966] volume 2, p. 91.
17. Popper [1961], pp. 135 – 6.
18. *Ibid.* p. 140.
19. See, for example, Popper [1966] volume 2, p. 91.
20. Popper suggests that there are good reasons for the belief that social sciences and social situations are less complicated than physics and physical situations on the grounds that humans act 'more or less rationally' (Popper [1961], p. 140).
21. See Popper [1979] especially chapters 3 and 4.
22. *Ibid.* p. 111, italics in original.
23. *Ibid.* p. 121.
24. *Ibid.* p. 161.
25. *Ibid.* p. 147.
26. *Ibid.* pp. 292ff.
27. *Ibid.* p. 294.
28. *Ibid.* p. 295.
29. Steven Lukes [1972] p. 86, makes the same point, although without elaboration.
30. See Popper [1967] and Popper [1961] p. 149f.
31. Popper [1967] p. 144, italics in original.
32. See Popper [1961] section 29.
33. Koertge [1974].
34. Popper [1967] p. 145.
35. *Ibid.* p. 144.
36. Koertge [1975] pp. 455 – 456.
37. Popper [1967] p. 149, my italics.
38. *Ibid.* p. 147. See also Popper [1979] pp. 178 – 179.
39. Koertge [1975].
40. For example, a plausible argument from a pre-Newtonian point of view denies the identification of physical space and the mathematical space of Euclidean geometry on the grounds that the former involves preferred directions, such as up and down, whilst the latter does not. An argument very like this is implicit in Aristotle's *On the Heavens*, II, 13, 295b 10ff. For other Aristotelian arguments against the possibility of a mathematical physics see his *Posterior Analytics*, 75b, 16 – 18, *Metaphysics* 995a, 15 – 20 and 997b 34 – 998a and *On the Heavens*, III, 8.
41. Lakatos [1978] p. 220.

6. POPPER AND LIBERALISM

Alan Ryan

It is clear to all readers of Popper's work that there is some sort of natural affinity between the account he gives of the rationality of science and his commitment to political liberalism. The object of this essay is to explore the nature of that affinity. The claims I make about it are initially very uncontentious, and hardly go beyond Popper's own words; I end, however, by making the more contentious claim that Popper's account of scientific rationality is itself in a broad sense political, and that what sustains his commitment to some awkward epistemological views is his liberalism. That is, it is not so much that his philosophy of science supports his liberalism as that it *expresses* it. This is not a claim which I imagine Popper himself would accept; indeed, I imagine that he would be extremely hostile to it. None the less, I should perhaps say at this point that it is not a claim made in any very hostile or critical spirit. Defences of liberalism, like defences of science, are almost doomed to waver somewhat between the thought that it is the *process* which justifies the result – a political decision or an accepted theory – and the thought that it is the *result* which justifies the process.

Since one of the things I wish to explore here is what kind of liberalism it is which Popper espouses, I shall follow his recommendation that we should avoid starting with elaborate definitions of our subject matter[1]. So I start by observing only that liberalism must, whatever else, place a high value on liberty and equality. It is true that Professor Dworkin has recently tried to persuade us that liberalism is concerned only with equality; but even he concedes that there will be liberal and non-liberal conceptions of equality, and that the liberal conception of equality will be characterized by its concern for individual autonomy[2]. So I think we shall come to no harm if we start from the

Currie, G. and Musgrave, A. (eds.), Popper and the Human Sciences.
 ISBN 90-247-2998-X

thought that liberals are concerned to achieve as much liberty and equality as may be.

Now, Popper remarks in his autobiography that he long ago came to the conclusion that although equality might be a good thing, it was excessively costly in terms of liberty[3]. But here, it is clear, the sort of equality he has in mind is something like equality of wealth or income. In *I he Open Society and Its Enemies*, Plato is attacked for his inegalitarian views, not, of course, because he defends inequality of wealth or income, for in these terms Plato's Guardians are worse off than the rest of the population of the republic, but because he defends a kind of political elitism which is at odds with the sort of equality to which Popper *is* committed[4]. That is, there must be at least an equality of basic rights; we may not wish to try to secure that the most energetic and the least energetic end up with the same amount of wealth or the same annual income, but we shall almost certainly want to start by giving everyone the same rights to acquire wealth or income, and we shall certainly want to give everyone the same political rights and immunities[5].

In accordance with another of Popper's own claims, we can agree that most liberals will see that these are values which we cannot pursue as 'absolutes'. If, say, freedom was understood as the absence of authoritative control over our behaviour, there is a strong case for saying that the creation of 'absolute' liberty by abolishing all such control would be self-destructive in virtue of permitting the strong to tyrannize the weak without restraint. Similarly, if we were to start with the notion of equality as absolute equality of wealth or income, we should soon find that any attempt to create *that* equality would be destructive of other sorts of equality – in particular, the need to police whatever system of achieving equality we dreamed up would destroy any sort of equality of power[6]. The idea that we can have absolute equality in the sense of equality in everything and in all ways is seemingly absurd, as is the idea that we can have absolute liberty; creating some equalities threatens others, just as leaving some liberties alone threatens the survival of others.

On this view, then, liberals will try to secure as much liberty and equality as possible; sometimes equality will strengthen liberty, and sometimes it will be in competition with it, just as sometimes liberty will sometimes reinforce equality and sometimes not. The kind of political equality which is enshrined in democratic procedures will, under favourable conditions, assist in the preservation of liberty by restraining rulers from trying to control every aspect of their subjects' lives, but under unfavourable conditions it may simply replace the tyranny of one man by the tyranny of the majority. There cannot be any conceptual demonstration that if only we pick the right conception of liberty and equality we can secure that there is no competition between them, and that we can have all we want of both – properly understood[7]. Or, more guardedly, any attempted demonstration of this compatibility is no use; we shall not know until after the event whether a given political choice embodied the right conception.

Given this minimal characterization of liberalism, it is easy to show how many of Popper's concerns have been central to the concerns of mainstream liberalism. Consider *The Poverty of Historicism*. The object of that work is to show that social prophecies of the kind embodied in Marx's social and political theory are no part of a rational programme of scientific research. The idea that it is the task of the social sciences to uncover the laws of history is attacked with a variety of weapons, but chiefly with the claim that the 'laws' on offer – say, Marx's 'law of increasing immiseration' – have not been laws at all, but trends[8]. They have not been properly framed universal hypothetical statements, but extrapolations of conjunctions of singular statements. The defender of Marx might argue that this is by no means a fatal objection. If a trend could be shown to be properly grounded in true laws and the appropriate initial conditions, it ought in principle to be possible to show that the trend would continue – in some appropriate fashion – for an indefinite period. There are, perhaps, two slightly different ways of explaining why this will not do. One might be to construe laws as Mill did, as asserting that whenever A, then a, unless something interferes; human affairs, of course, amount to a continuous attempt to interfere with what would have happened but for the interference, so there will be very little predictive power to be had from laws of this sort in social matters. Of course, we can sometimes predict how people will interfere, and make allowances for it, but we cannot do so at anything beyond short range. A man setting a booby trap engages in *ad hoc* predictions about how people will try to interfere with the workings of his device; but he cannot hope to predict how long it will take for them to get wise to everything he does – all he knows is that since they will be trying very hard, he himself had better keep innovating. A second way of putting the point, perhaps truer to Popper's own formulation, is to say that a law of the form $(x)(Fx \rightarrow Gx)$ may hold *so long as* people do not know of it; once they know of it, or perhaps even once they know of something else on which this law's holding depends, the law becomes false[9]. Again, the man setting the booby trap illustrates the point; the generalizations about the way people set out defusing a device on which he has hitherto depended turn out not to hold once their knowledge increases. The little semi-formal demonstration of the impossibility of determinism which prefaces later editions of *The Poverty of Historicism* makes the central point: we cannot predict what we are going to discover in future, but we do know that what we shall do depends on those discoveries[10].

It is in tracing the bearing of these arguments upon traditional conceptions of liberalism that we begin to see what sort of liberal Popper is. For, in the liberal tradition, we may discern two distinctive approaches, one most clearly espoused by Mill, one most clearly espoused by Kant; and these would yield rather different interpretations of why the arguments of *The Poverty of Historicism* matter so much. Suppose there were some people, social scientific experts, who *did* possess the sort of far-ranging knowledge of society's operations which we currently do not suppose they could actually have. What authority would this give them? On the difference between Kant's reply to this question and Mill's reply to the same question hang the two different liberalisms at stake here.

Kant's answer is unequivocal; such knowledge would not give its possessors authority over us. Their position vis à vis us is rather like that of a 'civilized' country proposing to colonize some 'primitive' people. However enlightened we may be, we have no right to colonize other countries[11]. However much our experts may know, they have no right to our obedience, unless we give them that right. In Kant's formulation of the point, he appeals to the idea of a social contract, not as a historical fact, since he knows as well as his successors that no such event took place, but as an Idea of Reason – a methodological device to remind us that we can only obey with a good conscience those authorities to whose rule we could have consented in the appropriate circumstances[12]. Popper's hostility to 'the myth of the origin' is such that it is hard to guess whether he would have much sympathy for what Kant was trying to do; at least, he would certainly want to rewrite Kant's appeal to a hypothetical contract in non-contractual terms. But, there is much in Popper to suggest that he would at any rate be sympathetic to the thought that each individual possesses a sort of moral inviolability which limits anyone else's claim to authority over him.

Now, Mill, who was perhaps a more consistent liberal than a utilitarian, can only accommodate something like Kant's restrictions on the grounds of authority by appealing to liberal ideals rather than individual rights, and at some cost to the supposed utilitarian foundations of his liberalism. That is, Mill's claim was that a sufficient disparity in knowledge between ruler and ruled did amount to a license for despotism[13]. Colonization was morally quite acceptable so long as its aim and effect were the improvement of the subject population. He was at least tempted by the appeal to expertise which ran through all of Comte's work. To get back to something like liberal principles, Mill resorted to history; once a society was 'civilized' its people were improvable by argument; once people were improvable by argument, the difference between more and less informed ceased to matter. All claims to knowledge are fallible – a view plainly congenial to Popper's criticalist account of science – and those who claim to know more than the rest of us are only to be trusted as far as they are willing to subject themselves to criticism from wherever it may come[14]. Whether Mill's appeal to fallibility is consistent with his generally inductivist approach to the growth of science is a moot point, but one we can leave on one side here.

More crucially, Mill had to appeal to the *ideals* of liberalism to defuse the claims of the knowledgeable. Liberty is one of the liberal's supreme values: he therefore minds very much about the *way* in which the less knowledgeable obey the more knowledgeable. They must be led with their eyes open; they must give the best assent they can, unforcedly, freely, and on the basis of such information and argument as they can be given. To this extent, Mill might even have held that Popper's appeal to 'piecemeal social engineering' was dangerously misleading as an account of the politics of the liberal state. For, it is characteristic of the inert, non-human material out of which bridges and automobiles are built that it has no view of its own, no values of its own, and could neither give nor withold its assent to anything proposed for it. But what makes us human is precisely that we do have our own values, our own views, and a taste for self-government rather than heteronomy[15].

Although I think that Popper's liberalism is much more nearly Kantian than Millian, it cannot be said that the texts are absolutely conclusive on their very faces. For Popper does not develop an elaborate account of rights in the way that Hayek, for instance, does do; even worse, from my point of view, Popper's clearest view is that the best defence of the political decencies is a kind of negative utilitarianism. That is, the constitutional problem is not Kant's problem – how to reconcile the freedom of each with the collective authority of all – but a problem more familiar in James Mill's *Essay on Government*. Since those who hold in their hands the power to do the good things we desire from government also hold in their hands the power to do evil, how can we design institutions to minimize the evil?[16] To the extent that this implies any very determinate position in political morality, it is that of a general humanitarianism, and a reminder that the miseries of the least significant member of society are miseries none the less, and therefore to be taken account of. At first sight, it is the negative case which is doing all the work: like Isaiah Berlin, Popper is frightened of theorists who believe that if only the right leader and the right doctrine coincided everything would go perfectly[17]. Like Berlin, he thinks there are no indisputable criteria for leadership, no special knowledge or special qualities of character to which they can lay claim. In this sense, liberalism is the product of scepticism even more than of a theory of inalienable rights, or of a commitment to the ideals of liberty. None the less, I think the beginnings of the positive case for liberalism are also to be found here – a positive case which is built up along with Popper's description of the practice of scientific research.

But one feature of the sceptical case also suggests the Kantian affiliation of Popper's politics. Popper insists on the logical distinction between facts and decisions – his version of Hume's famous distinction between *is* and *ought*[18]. This makes him more nearly Kantian than Millian on the connection between knowledge and authority. Mill seems to envisage superior minds knowing more than the rest of us not only about economics and sociology but also about what constitutes true elevation of character and the like. He is at least insecure about whether there can or cannot be moral experts, and this has given his critics a field day, with the more energetic claiming that his liberalism was the defence of an authoritarian secular enlightenment[19]. Kant, however, had no doubt that moral ideals were something which each person had to accept for himself. Technical advice could be proferred *ad lib*, and if we were sensible we would take the best advice we could get. But it was tyranny of the worst kind to force others to adopt our conception of their good, or force upon them our moral ideals.

It is central to Kant's thought that the state enforces external rules, which can be obeyed without further moral consideration, and that the authority of the state is limited to enforcing these. It is why property rights matter so much to Kant; through property rights and their cognates, we can deal with each other coercively but at arms length[20]. I promise you my bicycle for your stamp collection; once you have handed over the stamps I must hand over the bike; if I do not, you may properly force me to. What you and the state may not do is demand that I hand over the bike in any particular state of mind. It is my external performance only which is subject to control. This marks a concern to distinguish what is public, non-moral, subject to law and external from what is private, internal and moral. There is no such concern in utilitarianism; one of the oddities of Mill's *Liberty*, indeed, is that there is no mention of the distinction between public and private.

If this is right, the Kantian flavour of Popper's liberalism is not accidental. Further evidence for this comes in the conjunction of his uncertain attitude to democracy and his negative utilitarianism. Popper's defence of democracy is odd in that it is not based on what you might call the positive moral attractions of the democratic formula. Against Plato, with his contempt for the ordinary man, and his mystical faith in the qualities of the Guardians, Popper certainly defends democracy, in the sense that he defends the rights of the common man to a say in the government of his country and attacks the Guardians' claims to any such moral and intellectual infallibility as their authority is suposed to rest on. Moreover, says Popper, if the common man does not have the power to throw out rulers he wishes to throw out, he will be mistreated; all experience shows the truth of Lord Acton's *dictum* about the corrupting effect of power[21].

The justice of this as an attack on Plato is, of course, open to dispute. If one makes the first concessionary step of agreeing that Plato's intellectual universe makes sense, then the claim that *experience* shows that power corrupts will not seem quite such a knock down argument; Plato's reply is that, of course, power has always corrupted in the past because philosophers have not been kings and kings have not been philosophers. To my mind, this response really shows something else – that Popper was right to launch a *wholesale* attack on Plato; the aim cannot be to show the internal incoherence of Plato's case, but to show that this whole approach to politics is misguided from the ground up. Then, the defence of democracy can be made in the casual way Pericles makes it in *The Peloponnesian War*[22]. In essence, it is the old claim that the wearer of the shoe knows whether it pinches; not everyone can aspire to formulate policy, but what everyone can do is say whether the policies dreamed up by others are painful in their consequences.

There is more to it than that, of course. For one thing, what Popper and Pericles are eager to defend, and what Plato hates, is variety, vivacity, experiments in living. What Plato denounces as mere ignorant and childish running about after one fancy after another, Popper and Pericles praise as an attractive and enjoyable feature of the life of democratic Athens. As a defence of *democracy* in the narrow sense of the rule that decisions should be taken by some sort of majority vote, this would all be irrelevant, of course; but Popper is not concerned to defend that rule, since he knows as well as the next man that it can produce ugly results and paradoxical ones. The defence is more nearly a defence of toleration, what you might call a democratic view of cultural and moral issues, to the effect that the ordinary man has a right to live as he pleases within the limits of the law, and so long as he neither disturbs the peace nor prevents others exercising a similar freedom[23].

If this is plausible, it reinforces the point that this is a Kantian liberalism, since Kant's defence of the individual's rights is couched in very much the same terms. It offers a limited defence of democracy in suggesting that individuals have a right to a say in the affairs of their country, and, more importantly, that they are not likely to preserve their other rights unless they have some such say.

Primarily, however, it is a defence of individual rights, and only secondarily a defence of a decision making formula – which is one reason why Popper is more of a constitutionalist than a democrat, just as Kant was. The thought that this is the right lineage of Popper's views is not unduly disturbed by observing that Popper and Pericles side with Mill in defending experiments in living[24]. For Kant was every bit as concerned as Mill to defend such experiments. Kant's conception of the hidden point of history, it may be recalled, is that the goal of history is the development of all human faculties to their highest point. Of course, there is no empirical proof that this is the point of history; it is another of the Ideas of Reason which we employ to make sense of human experiences. We might be anxious about the difference between Mill's defence of variety for its own sake, and Kant's notion of a historical task; but, of course, it is axiomatic in Kant's system that we have a right to harmless enjoyments – we would not have assented to a government which could curb them – and it is implicit in his slightly awkward relation to his scholastic predecessors that he believes that the universe can only contain as much good as possible if it is as varied as possible too.

But it is at this point that Kant's liberalism and Popper's view of science begin to touch. Kant's defence of an evolutionary but non-utopian picture of human history partly rests on the claim that out of the crooked timber of humanity no straight thing is made; but mostly, it rests on the view that nature sets us a series of problems, the solution to which sets us new problems, and so indefinitely on. Kant's view that 'progress' is the goal of history is a sort of transcendental hypothesis, in something of the same way that Popper's conviction that increasing verisimilitude is the goal of science is a transcendental hypothesis. And one implication of this is that each individual who shares in the growth of knowledge or the development of human capacities may contribute only a little to the development of the whole species, but can, none the less, find a meaning to his life in so doing[25]. One ought not to be unduly surprised that two thinkers who adamantly reject the 'bucket' theory of the mind, and insist on the active role of the intelligence in interpreting experience should share more than epistemological allegiances.

If we place experiment, social progress, and fallibility at the centre of our politics, we have good reason to support a strenuous constitutionalism, which guarantees each man the maximum liberty consistent with the liberties of others. Democracy – in some sense other than conferring absolute power upon a majority – is one expression of this sort of constitutionalism, since anything other than equal political rights seems to flout the view that everybody has the same claim as everybody else not to be under anyone else's control without good reason. There is, of course, a more obviously utilitarian case to be made for democracy too, namely, that unless people can voice their views, even a well-meaning government may be ineffective for lack of information about what the people want. But we must recall Popper's view that the problem of government is not to secure that good men have the information to do good, but to secure that the non-rulers can throw the rascals out. Democracy, conceived as a device for the circulation of elites, is an answer both to the question of how to select competent rulers – by getting rid of the less competent ones peacefully – and to the question of how to retain power to throw the rascals out[26]. The reason why I mention the consistency of this defence of democracy with the 'elite theories' of writers like Pareto and Mosca is to further draw out the affinity between Popper's conception of democracy and his conception of science. In science, the essential is that no hypothesis can be protected by an appeal to authority. It does not follow that every scientist is equally good at testing old hypotheses or equally imaginative in suggesting new ones – let alone at carrying out tests of the new ones. What does follow is that the merits of a hypothesis are a matter of what happens to it under test, not a matter of who suggests it. In a democracy, not all ideas are equally good, and many policies are quite silly; none the less, we ought not to restrict the right to propose policies, and we must always remember that the merits of a policy are independent of the social or academic status of the proposer of it.

The way Popper's defensive view of democracy and constitution building place him in the Kantian tradition emerges even more clearly when we turn to his negative utilitarianism. His defence of this ethic is not elaborate or sophisticated – he has nothing much to say to the observation that a moral theory devoted to misery minimization would prescribe an unexpected and painless death for the whole of the human race. But this, too, is significant. Mill, whose *Liberty* claims in passing that governments do best when they stick to harm prevention, is aware that his official theory of government holds that governments ought to do whatever maximizes the general welfare; he sees that he needs to claim that liberty is an essential element in happiness, and that misery is more readily identified and more easily dealt with, if he is to square negative utilitarianism in practice with positive utilitarianism in theory[27]. Hence the complexities in his case which have irritated and excited critics for a hundred and thirty years.

Another view of the matter achieves the same results less awkwardly. Kant held that we simply had no business *making* other people happy. Governments might hinder hindrances to the search for the good life; but that was all[28]. Of course, this all becomes very much less simple when we probe it, and the line between hindering hindrances and positively promoting happiness in a paternalist fashion is harder to draw than it looks. All the same, the point of the doctrine is clear enough. Where Mill starts from the view that governments may legitimately do anything which promotes the general welfare, and then erodes that view in order to find room for individual liberty, Kant credits us with a right of self-defence, and allows governments to do only those things which help us to help ourselves without attacking others.

I do not wish to exaggerate the definiteness of Popper's case here, nor the clarity of its lineage. Popper is plainly at one with Mill in thinking that one reason for pursuing a negative utilitarianism is shortage of information. We are much more readily able to discover what makes people unhappy than what makes them happy, and on the whole what makes people unhappy is more uniform than what makes them happy; it is, also, quite often easier to cure. (This is, of course, very far from being universally true, and it is simple enough to think of exceptions.) Painters, poets, and pianists all suffer from cold and hunger, but a piano is no use to a painter, nor an easel to a poet. The state may properly engage in the provision of a range of basic welfare services, but is ill advised to do much more. It is not a major concern of Popper's to draw sharp lines around the proper tasks of the welfare state, and it would be wrong to try to extract too much from his argument[29]. All the same, it is perhaps worth noticing that the slightly blurred argument here, and the uncertainty about quite what tradition the argument belongs to, is characteristic of writers who have made much more of it than Popper. Hayek, for instance, sometimes employs utilitarian arguments against welfare state utilitarianism – or welfare state 'liberalism' in the American sense of the term; and sometimes, he employs arguments of a more strictly rights-based kind[30]. On one view, the ambituous state fails in its purposes because its information is inadequate and its techniques clumsy; on the other, positive steps to maximise social welfare involve illegitimate coercion in forcing members of society to part with goods and services they do not choose to part with, and for purposes other than self-defence.

I conclude, then, that Popper's liberalism is more nearly in a Kantian than a Millian mould, that it is a genuine liberalism in defending constitutionalism above a populist form of democracy, and that even where it is eclectic in its sources, it is classical liberalism in its concern for individual inviolability[31]. I turn finally to the task of showing that the connection between Popper's defence of liberalism and his view of science is partly a matter of instrumental and sociological arguments, but in part is simply a matter of pointing out the liberal virtues of the ideal scientific community. The instrumental case is relatively straightforward. Science seems to entail an open society, and an open society is indispensable to science. That is, if we manage to establish the scientific enterprise in any solid form, its achievements are sure to have an impact on society at large. We cannot believe one thing at work and another at home; we cannot study biology in the lab and heal ourselves with witchcraft at home. The degree to which conceptual incoherence is cheerfully tolerated by many people may be much greater than the out and out rationalist would like to believe, and the contents of our minds are not entirely transparent to us – whence, of course, Popper's own insistence on the need to *institutionalize* the pressure to think clearly and articulately. All the same, there will be a steady pressure to square the beliefs of society at large with the findings of science. This, however, is just what makes an open society: beliefs are not locked solid, rendered immune from investigation and challenge. If we live in a society with a flourishing scientific community, we shall find ourselves in a society with a pressure towards openness[32].

This, again, is a view which Popper seems to share with both Kant and Mill, and it is, of course, a contestable view; we are familiar with colleagues who seem to subscribe to one cosmology in the laboratory on weekdays and to another in church on Sundays, and no doubt there were physicists who were committed Nazis and no worse physicists for it. My own view is that a closed totalitarian society would stagnate scientifically after two generations, and would never do better than replicate work done elsewhere – but that is a hunch rather than a testable sociological hypothesis. What one can surely say is that no closed society can allow its scientists to chase hunches wherever they may lead; if the license to hunt is implicit in science, closed societies cannot practice science. Conversely, it is at least arguable that science needs an open society as background. The willingness to challenge established theories, the ability to go out on a limb without undue anxiety, the combination of eager questioning and patience to wait for answers, all seem to require the social training provided in a liberal society. The scientific community is one full of people who have passionately held convictions, but who need to resist the temptation to close ranks or shut up dissenters; the better the training in self-restraint provided elsewhere, the better the science. Again, the argument is contestable. Writers like Kuhn, who emphasize the closed and authoritarian nature of scientific communities cannot be lightly dismissed[33]. The violence of Popper's reaction to Kuhn suggests that he cannot bear to entertain the thought that liberalism and science may be systematic enemies rather than allies[34]

Kuhn, of course, represents himself as a modified disciple of Popper, but the point is still the same. Even if Kuhn's claim is only that scientists need to organize themselves in a more hierarchical and authoritarian fashion than Popper supposes if they are to carry out the eminently Popperian task of rigorously testing their conjectures about nature's laws, science would cease to be the natural ally of liberalism. Kuhn emphasises that among scientists ideas are not accepted regardless of who puts them forward; only some people are authorized to change the current orthodoxy, and then in orthodoxy conserving ways. If that is inescapable, we must accept that the habits of free speech and open discussion which characterize liberalism will be irrelevant to science. Conversely, if great stress is laid on the difference between what scientific claims mean to professional scientists and what they may fail to mean to laymen, the idea that there will be a sort of leakage from science to society at large which forces laymen to adopt new ideas and new ways of thinking will also be false.

In terms of the argument between Popper and Kuhn, one could pursue a conciliatory course, by pointing out that Popper stresses what Kuhn never denies – that nature tests our hypotheses, whoever devises them – and that Kuhn stresses what Popper never denies – that scientists need tenacity as well as a willingness to change their minds, that some attempts to protect hypotheses against premature falsification are justified, and that a scientific community needs some taken for granted ways of deciding where to direct its efforts[35]. Still, the story I wish to advance is at odds with this conciliatory intention. It starts with a familiar epistemological point. Many of Popper's critics have complained that once Popper's account of the process of conjecture and refutation is deprived of any residual empiricist or positivist elements, Popper can no longer tell us what hypotheses are tested *against*. Popper insists on a correspondence theory of truth, and also insists that there can be no statements which are simply known to be true in virtue of corresponding to the facts. The earliest version of Popper's account of science seemed to suggest that the assymetry between justification and falsification was important because we could know beyond doubt that some singular proposition was true, and use this as bedrock in testing general claims. This view is now said not to be Popper's own. All statements are revisable; all statements are interpretive, and general statements may be saved by numerous different strategies, some of which are quite justified.

The claimed virtues of science now become rather harder to elucidate, because the idea that scientific progress amounts to increasing verisimilitude is now more than ever an Idea of Reason in the Kantian sense, and the way is open for a much more 'sociological' account of science. That is, once it is accepted that we have a great deal of choice about what statements to accept or reject as part of the corpus of science, the pressure is on us to give an account of what amounts to good and bad choosing. The 'scientificity' of scientific doctrines becomes a matter of the process by which they are chosen rather than a simple matter of their factual truth. The obvious instrumental view that the goodness of the process of choice is simply a matter of the effectiveness with which the process eliminates false beliefs and replaces them with truer ones cannot be sustained in any simple way, once all ties with empiricism have been cut. At best, the claim that what the progress of science achieves is theories which are increasingly true to the facts becomes a metaphysical hypothesis, a heuristic maxim: 'act as if it were true that theories mirror facts'. At worst, Popper finds himself in the company of critical theorists such as Habermas, whose notion of truth is parasitic on what people would agree to say under conditions of freedom to say what they liked, equal ability to contribute to the consensus on what to say, and some sort of pressure towards reaching such a consensus[36]. It is going too far to suggest that Popper has arrived at the point of saying that the discoveries of science just are what the ideally liberal scientific community agrees to say they are, but it is not an exaggeration to suggest that he is under the same pressures in that direction as writers like Habermas have been.

I will not try to show what the consequences of this might be for Popper's defence of objectivity, though I cannot forbear remarking that the sort of tension I have pointed to seems to account for the critical anxiety which Popper's appeal to 'World 3' induces. I shall end by emphasising what I take it to show. It is not just that Popper espouses a rather Kantian liberalism; and it is not just that he thinks that the scientific frame of mind nourishes, and is nourished by, an open, liberal society. It is that the same Kantian picture of what it is like to try to make sense of the world which permeates his philosophy of science means that the bedrock reason for approving of science as an activity is that it is intellectual enquiry carried on by liberal means. To put it another way, it is the politics of scientific enquiry which makes it admirable, and the defence of political liberalism is not so much supported by an appeal to science as simply another part of a seamless web. This conclusion remains an uncomfortable one to square with Popper's hostility to what he thinks of as subjectivism, and I do not seek to disguise that. I will, however, say that it is not uncongenial to Kant's defence of enlightenment, with its cry of 'sapere aude', and not uncongenial to Popper's own defense of enlightenment, and his hostility to the irrationalism of Wittgenstein's later philosophy, with its defence of forms of life, and its claims for the inescapability of current and local habits of belief[37].

Postscript

I am grateful to Bryan Magee, John Watkins, and seminars at Oxford and Reading for their reactions to earlier versions of this essay. In two respects I feel that I have not been able to do justice to them. Professor Watkins thinks Popper mistaken in relinquishing the 'empiricist' elements in falsificationist methodology, and thus that I follow too readily down a road which leads nowhere. My only response is that I am impressed by the similarities between Popper's problems and those of Kant, and think the comparison illuminating. I am inclined to share Watkins's doubts about Popper's anti-empiricism – but that would have been a different essay. My question here was not so much whether that anti-empiricism was right, but whether the connections I suggest between Popper's philosophy of science and his political liberalism would hold up *if* (or *even if*) his anti-empiricism were right. Christie Davies (after the Reading seminar) suggested that I should have been sharper about the distinction between a liberalism founded on humanitarian concern and one founded on rights. He instances debates over abortion as occasions where one kind of liberalism is worried by the conflict between the rights of the mother and those of the unborn child, where the other worries only whether humanity to unwilling mothers carries unacceptable costs in other directions. I agree, of course, that these different forms of political argument have characterized debate in Britain and the United States. But Popper is not a natural rights theorist nor simply a rule of thumb utilitarian, so too sharp an insistence on the distinction would have meant pressing his writings for answers to questions they do not raise. I am inclined to say that, like Kant, Popper defends legal, constitutionally guaranteed rights, not because we just do have (natural) rights, but because they are the legal expression of something like Kant's principle of the inviolability of the individual. I must, however, admit that I am not sure that I could infer from what Popper has written on particular issues what he might say on those about which he has not written.

Notes

1. Popper [1962] volume 2, p. 17.
2. Dworkin [1978] pp. 127ff.
3. Popper [1976] p. 36.
4. Popper [1962] volume 1, pp. 94ff.
5. *Ibid.* volume 1, pp. 256–7.

6. Popper [1976] pp. 36 – 7.
7. Cf Dworkin [1978] pp. 123 – 6.
8. Popper [1961] pp. 115 – 6.
9. *Ibid*. pp. vi – vii.
10. *Ibid*. pp. v – vi.
11. Reiss [1970] p. 21.
12. *Ibid*. p. 79.
13. J.S. Mill [1914] p. 73.
14. *Ibid*. pp. 80 – 81.
15. *Ibid*. pp. 114ff.
16. James Mill [1957] p. 50.
17. Popper [1962] volume 1, p. 120.
18. *Ibid*. volume 1, pp. 60 – 61.
19. Cowling [1962].
20. Reiss [1970] p. 135.
21. Popper [1962] volume 1, pp. 136 – 7.
22. *Ibid*. volume 1, p. 7.
23. *Ibid*. volume 1, pp. 186 – 7.
24. J.S. Mill [1914] p. 115.
25. Reiss [1970] pp. 42 – 3.
26. Popper [1962] volume 1, pp. 124 – 5.
27. J.S. Mill [1914] pp. 132 – 3.
28. Reiss [1970] p. 134.
29. Parekh [1982] pp. 146 – 53.
30. von Hayek [1944].
31. Popper [1962] volume 1, pp. 99 – 104.
32. *Ibid*. volume 2, pp. 217 – 22.
33. Kuhn [1962].
34. Popper [1970] pp. 51 – 9.
35. Kuhn [1962] pp. 231ff.
36. Habermas [1970] pp. 360ff.
37. Reiss [1970] p. 54.

7. MAKING SENSE OF CRITICAL DUALISM

Jeremy Waldron

The theory called '*critical dualism*' is one of the most important, but far from the clearest, parts of Karl Popper's conception of the way history and the human sciences are related to morals and politics[1]. It is a theory about the relation between the facts or factual laws which the human sciences might discover and the norms or decisions which we might adopt in the face of those facts. Norms and decisions, Popper asserts, cannot be *derived* from facts; no belief that a certain particular or law-like state of affairs prevails in the world can *commit* us logically to any decision regarding our conduct. It is for us to decide what to do with regard to any fact that we take to be the case, to alter it if it is alterable, or to oppose attempts by others to alter it.

The importance of this theory to Popper's conception of the open society is obvious. If we say it is up to us what to do about natural and social facts, then we foster and encourage a critical rather than a submissive or magical attitude to the institutions that surround us[2]. And, of course, critical dualism is intimately related to Popper's attack on historicism:

> Neither nature nor history can tell us what we ought to do. Facts, whether those of nature or those of history, cannot make the decision for us, they cannot determine the ends we are going to choose .... Facts as such have no meaning; they can gain it only through our decisions. Historicism is only one of many attempts to get over this dualism; it is born of fear, for it shrinks from realizing that we bear the ultimate responsibility even for the standards we choose (Popper [1966] volume 2, pp. 278 – 279).

Stated in this way, with its emphasis on our responsibility for the choice of norms and standards, and the freedom of commitment we have in the face of any facts, Popper's critical dualism sounds a lot like

Currie, G. and Musgrave, A. (eds.), Popper and the Human Sciences.
 ISBN 90-247-2998-X

the modern meta-ethical thesis of the 'fact-value gap'. In modern moral philosophy, that thesis is usually associated with some form of ethical scepticism: the reason why statements of value cannot be derived from statements of fact is that there are no values in the world to be connected with facts in the way that a derivation of the former from the latter would require. Instead, it is argued, value-statements are nothing more than expressions of our attitudes or reactions to the world. From an objective point of view, facts and our subjective response to them are all that there is. The realist picture, of a world containing values, norms and standards in itself, of which our normative commitments are more or less adequate representations, is rejected[3].

Although Popper's theory sounds similar to this modern ethical scepticism, there is an immediate difficulty in identifying it as such. For Popper describes his theory not as a scepticism or an anti-realism about norms and values, but as a *dualism*.

Now talk of dualism is usually realist talk. In the context of the mind-body problem, for example, dualism is the theory that minds, or at least mental events and states of consciousness, exist as an order of reality distinct from physical things and events, such as brains and brain processes. Certainly, dualism rejects any attempt to 'reduce' the mental to the physical or to characterize the former as a 'logical construct' out of the latter; in this respect, then, it has *something* in common with a sceptical or anti-realist 'eliminative' materialism[4]. But it maintains that the irreducibility is evidence of the distinct reality of mental phenomena, not of their ontological dispensibility.

Popper himself is at least a dualist in this sense – 'at least', because in fact he believes that there are not just two but *three* 'worlds' or orders of reality: World 1, the world of physical things and events; World 2, the dualist's world of mental and phenomenal states; and World 3, the world of 'objective contents of thought', such as propositions, theories, books, symphonies, and so on[5]. This is a very controversial metaphysic, but it is not in question here; I have introduced it purely to raise the issue of the nature of Popper's 'critical dualism'. Is the dualism of facts and norms supposed to be like the dualism of the mental and the physical? Does it assert the reality of norms and values in the way that mind-body dualism asserts the irreducible reality of states of consciousness? If so, then Popper's meta-ethic is far from the moral scepticism with which we earlier compared it. It is more like the so-called 'intuitionism' of Prichard or Moore[6]. Moore, of course, takes the credit for philosophical interest this century in the 'naturalistic fallacy' – the fallacy of supposing that goodness is identical with, or definable in terms of, any natural fact or quality. But Moore draws no sceptical conclusions from this irreducibility: on the contrary, he asserts that goodness is a non-natural quality of things which is as much a real feature of the world as the qualities studied by chemists, physicists, and psychologists[7]. So, for Moore, the dualism between natural fact and value is exactly on a par with psycho-physical dualism. (Indeed, this is a parallel that Moore explicitly draws[8].)

Is it plausible to attribute a view like this to Popper? Certainly, this theory – that values and norms are *sui generis* features of the world, not indeed reducible to the 'natural' and social facts discovered by the sciences, but no less real for that – is consonant with some of what Popper says about ethics. In his initial exposition of critical dualism in *The Open Society*, Popper is anxious at all times to avoid any association of his views with relativism or nihilism. He adamantly rejects any inference from dualism to the claim that norms and decisions are 'merely arbitrary'[9]. In an important addendum to *The Open Society*, Popper goes so far as to suggest that the realm of values admits of a regulative idea of *rightness* or *validity* which performs a function analogous to that of *truth-as-correspondence* in the realm of facts[10]. He suggests, too, that in the realm of values, 'as in the realm of facts, we can make *discoveries*':

> That cruelty is always 'bad'; that it should always be avoided where possible; that the golden rule is a good standard which can perhaps even be improved by doing into others, wherever possible, as they want to be done by: these are elementary and extremely important examples of discoveries in the realm of standards.
> (*Ibid*, volume 2, p. 386).

Again, though Popper says little about contemporary work in meta-ethics (I shall discuss this feature of his writing later), what he does say reveals a hostility towards the anti-realism of logical positivist and emotivist theories. Rejecting the idea that norms are meaningless or that they are mere expressions of emotion, Popper ventures the opinion

> ... that the reluctance to admit that norms are something important and irreducible is one of the main sources of the intellectual and other weaknesses of the more 'progressive' circles in our present time.
> (*Ibid*. volume 1, p. 234).

These are all aspects of Popper's meta-ethic which, on a first reading at least, might prompt us to identify it as a sort of Moore-style non-naturalist cognitivism.

But how would such an interpretation handle what one might call the *libertarian* side of Popper's position: his insistence that norms are man-made, that they enter the world only when human beings begin to make critical moral judgements, and that we alone are responsible for all the norms and standards that exist? There are several points to make here.

First, there is a tendency for Popper's meta-ethics to get mixed up with another theory that he holds – indeterminism or free-will. Sometimes, indeed, Popper formulates his critical dualism as if it were an indeterministic thesis: for example, 'Facts ... cannot *determine* the ends we are going to choose'[11]. But it is easy to see that the two positions – indeterminism with respect to choice of values, and the non-derivability of values from facts – are distinct. A determinist may be a moral non-cognitivist: he may believe that the values that we choose cannot be logically derived from any statement of the facts which causally determine the choice of them. And, of course, many of those who believe in some version of ethical naturalism have also believed that humans are free to choose whether or not to fly in the face of the values and standards they take to be implicit in nature. In general, you can be as realistic as you like about values without giving up the thesis that the facts – even the moral facts – cannot determine our decisions and our ends for us. For a moral realist, the most that the facts determine is whether our free decisions and freely chosen standards are right or wrong, true or false[12].

Secondly, interpretation of Popper's position at this point is made difficult by his uncertainty as to what his dualism is a dualism between. It is introduced initially in *The Open Society* as a dualism between facts and *norms*[13]. Now, facts remain on one side of the dualism throughout Popper's writings on the subject, but he never seems to have been happy with his specification of what is on the other side. Almost immediately, in *The Open Society*, there is a shift from norms to *decisions*[14]. In a footnote which was added later, Popper takes up the suggestion that he talk about *proposals* instead of decisions[15]. In the 1961 addendum, proposals are joined, if not usurped, by *policies*, and there is dark talk too of *standards*[16]. Finally, in his more recent 'Replies to my Critics', Popper claims it was a mistake to put '... "proposals for a policy" on a level with "decisions" ', and he seems to want to go back more or less to where he began with a dualism of facts and decisions[17].

This vacillation may be just a matter of terminology, as Popper suggests at one point[18]; if so, there is no reason to dwell on it. But metaethics is a treacherous area of philosophy, and all too often terminological slides conceal substantial equivocation. I think it is worth pursuing the matter a little further.

Consider a particular decision: for example, my decision not to harm other people. When somebody claims to be talking about my decision, there seem to be two things he might be talking about: (a) he may be talking about the mental event which consists in my actual making of the decision; or (b) he may be talking about the objective content of the decision that I make. In sense (a), he is talking about an item in Popper's World 2. Since decisions in this sense are facts about people, clearly this is *not* the sense in which decisions are to be *contrasted* with facts[19]. (There might be a contrast between facts of this sort and, say, facts of World 1: but this amounts, as we have seen, to mind-body dualism and perhaps to indeterminism.) Decisions in sense (b) are World 3 items. They might be referred to in some such exchanges as this: 'So, what was his decision, then?' – 'Not to harm other people.' – 'That's the same decision as mine.' Popper provides a useful analogy to illuminate the distinction between senses (a) and (b). The decision in sense (b) stands in the same relation to the decision in sense (a) as the proposition 'Napoleon died on St. Helena' stands to the event of a historian's claiming that Napoleon died on St. Helena[20].

But this analogy suggests that there might be a further sense, (c), to be considered. For the proposition 'Napoleon died on St. Helena' is to be distinguished, not only from the event of a historian's claiming that Napoleon died on St. Helena, but also from the event which the proposition describes, and which, if it is true, makes it true: viz., Napoleon's death on St. Helena. Analogously, we might want to distinguish my decision in sense (b), not only from the event of decision-making in sense (a), but also from (c), the state of affairs (that people are not to be harmed), which makes the decision in sense (b) right or true or valid.

Admittedly, it is odd to regard sense (c) as a sense of 'decision'. Normally, if it is decisions we are talking about, the distinction between senses (a) and (b) is sufficient. (The same seems to be true of proposals and policies.) But if we are talking about *norms*, or *values* or *standards*, then, if we take moral realism seriously as a philosophical theory, we need something like sense (c) as well. Thus, for the norm

'Thou shalt not kill', we will want to distinguish between: (a) the norm as a positive social fact;[21] (b) the objective content of the norm; and (c) the state of affairs in moral reality – that people are not to be killed – which makes the norm true or valid. Of course, moral sceptics will deny that there *is* a norm in sense (c); nevertheless they need sense (c) to make sense of the (false) claims that they think the realist is making[22].

Now, when Popper talks of a dualism of facts and norms, we know that it is not norms in sense (a) that he is interested in. But is he interested in sense (b) or (c)? No clear answer to this question emerges from Popper's writings. But it makes an important difference. For it is possible that Popper is realist about norms in sense (c), and that all his talk about the man-made character of norms, and about our responsibility for them, refers to norms in sense (b) only. As the objective contents of certain mental states, norms and decisions in sense (b) would be residents of Popper's World 3. According to Popper, World 3 is 'the world of the products of the human mind'[23]. The items in it 'originate as a product of human activity'[24] and, although they are capable, once created, of taking on an objective life of their own, still they remain susceptible to change and modification by humans. Thus the thesis that norms are man-made and that we bear ultimate responsibility for them is quite compatible with the thesis that norms and values exist in the world and make our moral judgements true or false, valid or invalid, whether we like it or not – provided that we distinguish norms in sense (b) from norms in sense (c).

Popper, then, might be a moral realist. True, the meta-ethical ideas in *The Open Society* were set out long before Popper articulated the metaphysical doctrine of World 3. But once that doctrine has been articulated, this seems a natural way to interpret what Popper says about norms in *The Open Society*.

But now notice two points about the position we have been attributing to Popper. First, if he is a moral realist and if he allows the three-level distinction which we have been discussing, then his claim that norms and decisions are man-made is no longer an *argument* for critical dualism, i.e., it is no longer an argument for the non-derivability of values from facts. For in *exactly* the sense in which norms are man-made (sense (b) above), theories and propositions about matters of fact are man-made too. A theory, no less than a norm, is a resident of Popper's World 3, and it establishes its residence there only as a result of human action. Popper concedes in a footnote that this is so, but he maintains that there is *still* a difference between norms and hypotheses, justifying his claim that there is a logical gulf between them. He argues that 'although the hypothesis is man-made, we may be unable to prevent its falsification', since, independently of our will, the facts may falsify the factual claims that we make; in the case of norms, however, it is for us, not only to create the norm, but to decide whether or not it is broken, for it is our decision whether or not to obey it[25]. If moral realism is true, this is a bad argument. Consider the difference between the *breach* and the *falsification* of a universal norm such as 'Thou shalt not kill'. The fact that a norm has been broken or disobeyed tells us nothing, according to the moral realist, about whether the state of affairs it purports to represent actually obtains. According to the moral realist, the relation between the norm and the state of affairs will be expressed in some such formula as this:

> 'Thou shalt not kill' is true if and only if no-one is to kill anybody.

The consequent of this biconditional may be true no matter how many murders take place. Conversely, it may be false even though everyone on the face of the earth obeys the Sixth Commandment. So norms may be on a par with hypotheses so far as truth and falsification are concerned: just as 'All ravens are black' is falsified by the existence of a single green raven so, according to a moral realist, 'Thou shalt not kill' is falsified by the existence of a single case where it is right to kill someone. This leaves obedience and disobedience to a norm as a *further* category – a matter of people's belief in its truth or validity and their attitude towards the state of affairs it represents. But there are analogues for all this in the case of hypotheses and the states of affairs that they describe.

So the first point about interpreting Popper as a moral realist is that it leaves him without a basis for his dualism. He can, of course, still maintain it; but it is no longer based, as he suggested it was, on any disanalogy between norms and factual propositions. Instead, it will have to be based on something like G.E. Moore's 'open question test'. But it is not clear that Popper wants to have any truck with that.

However, a second, and far more important point about the position we have been attributing to Popper is this. All the stuff about norms and decisions, in sense (b), being World 3 items and, as such, objects of critical discussion and rational argument, could be maintained *even though the whole idea of moral realism is dropped*. Thus, suppose Popper were to join such sceptions as John Mackie or Simon Blackburn in denying that there are any norms, values and decisions in sense (c)[26]; the distinction between senses (a) and (b), and all that that entails, would still remain. In the rest of this paper, I shall defend the claim that this, in fact, *is* Popper's position in meta-ethics: that he is, despite occasional suggestions to the contrary in *The Open Society* a moral sceptic (as moral scepticism is nowadays understood), and that this position is, even to a rationalist, an attractive and defensible one, despite the misgivings about its implications which prevented Popper from embracing it explicitly.

It needs little argument to show that norms, values and decisions can be part of the fabric of World 3, even though there is no further moral reality to correspond to them or make them true or valid. Although many of the items in World 3 are representational, i.e. purport to represent some state of affairs beyond themselves, this is by no means true of them all. Descriptive propositions and scientific hypotheses are representational, but, among the other World 3 items that Popper mentions, symphonies, numbers and universals are striking example of ones which are not. Scepticism, then, about there being anything for norms and decisions to represent or correspond to, does not commit us to withdrawing them from World 3.

It is, however, true that most moral sceptics are *emotivists* (if that term is understood in a suitably broad sense). They believe that moral statements involve nothing more than the expression of subjective attitudes or responses. (This is not quite true: most emotivists now concede, under pressure from writers like Philippa Foot, that there is a limited descriptive content to many moral concepts, but they insist that their distinctively moral core is expressive rather than descriptive[27].) But Popper associates emotivism with 'psychologism':

> The followers of 'psychologism', on the other hand, will try to explain imperatives as expressions of emotions, norms as habits, and standards as points of view (*Ibid.* volume 1, p. 234).

And, of course, psychologism is supposed to be anathema to all the devotees of World 3 objectivism. But, even if emotivism *is* psychologistic, I do not think it is much of a threat.

In the 1961 addendum to *The Open Society*, Popper develops a series of arguments against what appears to be an emotivist position:

> ... the relationship of evaluation (of some questionable fact by some adopted or rejected standard) is, logically considered, totally different from a person's psychological reaction (which is not a standard but a fact), of like or dislike, to the fact in question or to the standard in question (*Ibid.* volume 2, p. 384).

Popper's argument to this conclusion is that, even if our likes and dislikes play a role in determining which standards we accept, nevertheless it is possible to judge or evaluate a fact by a standard we do *not* accept; so evaluation cannot be a matter of expressing likes or dislikes. This argument is fallacious. With regard to any fact *F* and standard *S*, it is necessary to distinguish between: (i) saying how *F* stands with respect to *S*, i.e. whether or not *F* complies with *S*; and (ii) evaluating *F* using *S*. Suppose *F* is the death of six million Jews and *S* the ideal of Aryan purity: then it is possible for me, as a liberal, to perform activity (i) but not activity (ii). I can talk about the Nazi standard and say whether the events of the holocaust live up to it, but I would never use this standard to *evaluate* the holocaust. My condemnation of it would be based on quite different standards. Popper's argument does establish something: that standards are World 3 objects, which can be talked about and discussed whether we like or accept them or not. But it does not establish that the use of a standard in evaluation is not a matter of expressing an attitude.

Popper also makes the point that evaluating a fact is 'far from being the same as asking whether we like it'[28]. This is absolutely correct. But it is no argument against emotivism unless we commit the howler of identifying emotivism (the theory that evaluation is the expression of an attitude) with subjective naturalism (the theory that evaluation is saying that one has an attitude).

Finally, Popper makes the point that 'our likes and dislikes are facts which can be evaluated like any other facts'[29]. Again, this is correct, but it does not show that evaluation is not a matter of expressing likes, dislikes or (as I prefer to say) attitudes. An emotivist account of Popper's insight here would go as follows. To have an attitude towards a state of affairs (say, the system of apartheid) is to respond to certain features of it in an affective and appetitive way. Now, we are sophisticated beings and capable of being quite self-conscious about this. We may criticize the attitudes or, more generally, the moral sensibilities (i.e. the dispositions to form attitudes) of those around us: we may think, for example, that someone's attitude of tolerance towards apartheid, expressed in his view that it is not an evil system, is a response only to certain features of apartheid (its anti-communism, for instance) and that he is blind to some of its other features (for instance, the suffering it causes). We may condemn his attitude and regard the sensibility that generates it as defective inasmuch as it involves some sort of failure of imagination, a failure to identify with the plight of those who suffer under apartheid. But if we couch this criticism in a general form we may then turn it against ourselves, as we consider the possibility that our own moral sensibility contains elements or deficiencies which we would not admire if we spotted them in others. Now, *of course*, it remains true that all this evaluation and self-criticism is nothing, in the last resort, but the expression of attitudes towards attitudes. But that need not worry us, for it threatens neither circularity nor dogmatism. The only constraint it imposes is this: that we cannot evaluate all our attitudes at once. But none of them need be given axiomatic status or put beyond criticism. Like the planks in Neurath's boat, being rebuilt at sea, holding some of them constant at any given time, we are able to evaluate and if need be replace each one in turn[30].

It *is* possible, then, for the attitudes which, the emotivist says, are expressed in moral statements to be the objects of argument and criticism. We can ask such things as: Is my attitude towards apartheid the same as yours? Is it better or worse than yours? Is it consistent with my other attitudes? How could it be improved? and so on. We can even make sense of Popper's talk of discoveries and progress in ethics. For example, our attitude to cruelty is not what it was; it has improved to the extent that most of us now agree that no end can justify the employment of means that are deliberately cruel. As Popper indicates, in making these discoveries, 'we have to lift ourselves by our own bootstraps'[31], but that does not mean that it is senseless to talk of discoveries or progress in ethics.

If all this is granted, but it is *still* maintained that attitudes are 'too psychological' to figure in World 3, then I have to confess to being unable to see any longer why World 3 status should be thought to matter[32].

It remains to deal with Popper's worry that, despite all this, moral scepticism leads in the direction of nihilistic or relativist views. These latter positions are notoriously ill-defined – that, for the people who hold them, is usually one of their main attractions! At one point, Popper simply defines *nihilism* as extreme moral scepticism[33]; but usually he takes it to be the view that nothing matters, or that life should be lived on the verge of destruction and nothingness, or some such juvenile barbarism. Now, to hold any of these views is to adopt certain attitudes towards life (e.g., attitudes of indifference to things most people regard as important, or attitudes of barbaric recklessness) and to reject others. But it is obvious that emotivism – the theory that making a moral judgement is expressing an attitude – does not commit us to any attitude in particular, let alone any of these. Most of the people whose moral practice emotivism purports to describe would regard nihilistic attitudes as evil or crazy. And with good reason.

Relativism is also initially identified by Popper with scepticism[34]; but it is mostly taken to be the view that the choice between values, norms and decisions is 'arbitrary'. The answer Popper gives to relativism in this sense is clear and correct, and could survive his adoption of moral scepticism[35]. To say that the choice between norms is arbitrary is to imply that one system of norms is as good as any other. But this is itself a normative or evaluative judgement (since it denies the claim that one of the systems is better than any of the others). Clearly, moral scepticism does not commit us to this view: scepticism is at most the theory that there are no real states of affairs in the world which make our norms and evaluations true; it is not the theory that there is a real state of affairs in the world which is the moral equality of all norms. It is not inconceivable that someone may adopt the attitude which relativism implicitly recommends. But there is no rational compulsion to do so, and most of us would condemn and criticize anyone who did.

Sometimes relativism is associated with the theory that whatever any person or society thinks is right *is right for them*, and that therefore it is a mistake to offer trans-personal or cross-cultural moral criticism. It is difficult even to formulate this position coherently, and certainly impossible to defend it. The conclusive argument against it was stated succinctly by A.J. Ayer:

> On my analysis, to say that something which somebody thinks is right really is right is to range oneself on his side, to adhere to that particular standpoint, and certainly I do not adhere to every standpoint whatsoever (Ayer [1954] p. 247).

Unfortunately, Popper misunderstood this point of Ayer's when it was put forward by Lord Boyle in a discussion of Popper's meta-ethics[36]. Popper's comment was as follows:

> ... this would make the Mafia adherent 'right', who ranges himself on the side of criminals, and adheres to that particular standpoint: he would be 'right' in the relativistic sense if we apply this sense of the term 'right' to him, or to those who 'unite their wills with those of the others' belonging to the Mafia (Popper [1974] p. 1158).

This is a dreadful confusion. What follows from Ayer's position is this: if the Mafia man ranges himself on the side of the criminals, then, to that extent, he shows that *he believes* their conduct to be right; conversely, if he says that what the criminals think is right really *is* right, then he ranges himself on the side of the criminals. But none of this commits *us* to the judgement that either the Mafia man or the criminals, thinking and ranging themselves as they do, is 'right' (in *any* sense of the term 'right'). *We* are not committed to the moral views of another person merely because we understand what it is for him to have and express those views; and it is the relativist's mistake, not Ayer's, to think otherwise.

To sum up. I have attempted, in this paper, to make sense of the position in meta-ethics which Popper calls *critical dualism*. We began with the hunch, inspired by the realistic connotations of the term 'dualism', that Popper is a moral realist; that, rather like G.E. Moore, he believed in the existence of a moral reality co-ordinate to but independent of the reality of social and natural facts. But we found this, in the end, less plausible as an interpretation of Popper's position than one which makes him *sceptical* about such a moral reality, and which associates his position with a mature and sophisticated emotivism. In the last few pages, I have attempted to show that such a position is consistent with all of Popper's important insights into the nature of morality: his belief that norms and values pertain to, but are not derivable from, the facts of natural of social life; his fallibilist conviction that no view (in ethics or anywhere else) is authoritative or beyond criticism; his faith in the possibility of moral progress and his use of some regulative idea analagous to truth to express the notion of such progress; his implacable opposition to all forms of irrationalism; and his theory that, at the end of the day, we, with our evaluations and commitments are the creators of, and are therefore responsible for, all the norms and values that there are.

Finally, some thoughts on the relation between Popper's work and mainstream moral philosophy. The task of interpreting critical dualism has not been made any easier by the almost complete lack of dialogue on these matters between Popper and those philosophers who have been at the forefront of recent meta-ethical discussion. As we have seen, there is scant reference in Popper's work to that of his contemporaries in the area, and, where their work is referred to, it is often misunderstood. It is true that meta-ethics was at a low ebb in the wake of logical positivism, and that many of the claims of the early emotivists were such as to give rise to the misapprehensions and misgivings which characterized Popper's response to them. But by 1961, when the addendum, 'Facts, Standards and Truth' was added to *The Open Society*, R.M. Hare's *Language of Morals* had been published for ten years, and the sophisticated form of emotivism espoused therein was dominating the debate in mainstream moral philosophy. As far as I can tell, nowhere in the addendum or anywhere else in Popper's writings, is there any mention of Hare or Hare's opponents. (This must be part of the explanation of the fact that, as Popper has recently complained, the addendum 'has been, so far as I know, overlooked by almost everybody[37].) Even more worrying is the almost complete absence of reference back, in Popper's writings on ethics, to philosophers such as Hume or Moore, whose positions are usually taken to be landmarks in the modern debate[38]. (Kant is a notable and important exception[39].)

On the other side, the neglect of Popper's meta-ethical writings by those involved in the mainstream of contemporary moral philosophy is equally culpable. There are occasional references to Popper's theory of scientific method: Hare, for example, relies on an important analogy between universal prescriptivism and falsifiability[40]. But, as far as I can see, there is no sustained discussion of Popper's critical dualism in any important work in meta-ethical theory published since the war[41]. Even in two recent monographs devoted to the exposition of Popper's philosophy, there is no discussion of his meta-ethics[42].

All this is regrettable for three reasons. First, however unpalatable the fact may be in Oxford, Popper continues to be regarded as a major figure by the non-philosophical world and by philosophers outside England and America. It is significant, for instance, that when Jurgen Habermas wants to talk about the doctrine of the 'fact-value gap', he cites Popper's work in this area as a primary source[43]. At a time when Anglo-American philosophy is desperate to re-establish its link with the Continental traditions, it is sad and ironic that it should ignore the work in ethics of one philosopher who continues to be taken seriously on both sides of the Channel.

Secondly, Popper's work in meta-ethics has always been located in the sort of wider philosophical framework that is increasingly being recognized as necessary if progress is to be made in this area. Most of us are now convinced that work in moral philosophy needs to be brought into relation, on the one hand, with the discussion of first-order normative and political questions – questions about justice, property, war, rights, and the state. And there is growing recognition that, on the other hand, moral philosophy must not be studied in isolation from wider epistemological considerations. The questions about truth, realism, meaning and knowledge that arise in meta-ethics are substantially the same that arise, for instance, with regard to science and mathematics. It would be a mistake to think we can be satisfied with answers to these questions in ethics without bringing them into systematic relation with the answers that are being given in the latter fields. Popper's work in meta-ethics is significant in both these respects, for it is integrated, on the one hand, with a theory of science, and, on the other, with an articulated political philosophy. In these respects, it represented an important return to the traditions of Hume and Kant at a time when much of what passed for philosophy was distinguished by its specialization.

Finally, we should bear in mind that philosophy, particularly as it is done today, is a co-operative enterprise. Even when they appear to be engaged in endless and internecine conflict, philosophers are building on one another's work; the discipline thrives on contact, exchange, understanding and criticism, and it is harmed and undermined if we drift into 'schools of thought' who will not talk to one another. This is as important in moral philosophy as anywhere. The terrain is treacherous and it is easy, as we have seen, for even straightforward positions to be dreadfully misunderstood. Little of what I have said in this paper is original or new. But I have tried to apply to the interpretation of Popper's meta-ethical theory some of the arguments and insights of mainstream thought in that area, from which it has been isolated too long.

Notes

1. Critical dualism is expounded in Popper [1966], mainly in chapter 5, but also in an addendum, 'Facts, Standards, and Truth: A Further Criticism of Relativism', added in 1961.
2. Popper [1966] volume 1, p. 57ff and volume 2, p. 399.
3. Such a scepticism or anti-realism is offered by J.L. Mackie in his [1977], especially chapter 1.
4. For eliminative materialism, see R. Rorty [1980] pp. 114ff.
5. This three-world pluralism is outlined and defended in Popper [1979] Chapters 3 and 4; and in his lecture 'Three Worlds' (Popper [1980]).
6. 'Intuitionism' is a popular misnomer as neither Moore nor Prichard accepts the epistemological theory of intuition: see, e.g. G.E. Moore [1903], preface and chapter 5, section 86. Nor for that matter does Popper: see his [1966] volume 2, pp. 15–16 and accompanying note, and p. 388ff.
7. See Moore [1903] chapter 2, sections 25–26.
8. *Ibid.* chapter 1, section 12; see also Moore [1909–10].
9. Popper [1966] volume 1, p. 61 and pp. 64–65.
10. *Ibid.* volume 2, p. 385.
11. *Ibid.* volume 2, p. 278.
12. I think Popper realizes this; see the addendum to his [1966] volume 2, p. 391.
13. *Ibid.* volume 1, p. 60.

14. *Ibid.* volume 1, p. 65.
15. *Ibid.* volume 1, pp. 234 – 235.
16. *Ibid.* volume 2, p. 383.
17. Popper [1947] p. 1155.
18. Popper [1966] volume 2, p. 384.
19. *Ibid.* volume 1, p. 63.
20. *Ibid*, volume 1, pp. 63 – 4.
21. The best account of what it is for a norm to exist as a social fact is still that given by H.L.A. Hart in his [1961] pp. 54ff.
22. Some moral philosophers deny that they *can* make sense of the realist's claims; for a discussion of this possibility, see Mackie [1977] pp. 20ff.
23. Popper [1980] p. 167.
24. Popper [1979] p. 159.
25. Popper [1960] volume 1, p. 239.
26. See Mackie [1977] chapter 1, and S. Blackburn [1971].
27. See, e.g. C. Stevenson [1937]. See also P. Foot, 'Moral Beliefs' and 'Moral Arguments' in her [1978]; see also R.M. Hare [1952] chapter 7, and Mackie [1977], p. 23.
28. Popper [1966] volume 2, p. 384.
29. *Ibid.* volume 2, p. 384.
30. In this paragraph and elsewhere, I am greatly indebted to Simon Blackburn's discussion of 'quasi-realism' in ethics: see his [1971], and especially his [1981] pp. 174ff.
31. Popper [1966] volume 2, p. 386.
32. If a less 'psychologistic' emotivism is required, Hare's theory of 'prescriptivism' may be referred to: see Hare [1952], especially chapter 1.
33. Popper [1966] volume 1, p. 82.
34. *Ibid.* volume 2, p. 369.
35. See *ibid.* volume 1, pp. 64 – 65.
36. E. Boyle [1974] p. 851.
37. Popper [1974] p. 170, n. 177.
38. Moore is mentioned briefly in a footnote, but dismissed for being too interested in the concept or essence of goodness. Hume's meta-ethical position is not mentioned at all.
39. For a discussion of the relation between Popper's position and Kant's, see B.T. Wilkins [1978] chapter 4.
40. R.M. Hare [1963] pp. 87 – 92; the only reference to Popper's meta-ethics in Hare's work is the briefest citation in his [1952] p. 30.
41. This is certainly true of the well-known monographs by Baier, Brandt, Foot, Frankena, Hare, Hudson, Mackie, Nowell-Smith, Richards, Toulmin, Urmson and Warnock. Chapter 3 of P. Winch's [1972] is an honourable exception.
42. B. Magee [1973] and A. O'Hear [1980].
43. J. Habermas [1973] p. 276.

8. BEYOND CULTURAL RELATIVISM

Noretta Koertge

1. *Introduction.* Any philosopher of science who attempts to do field-work amongst the tribe of anthropologists is in for severe culture shock. I'd like to illustrate with two ethnographic examples.

Anecdote 1: At a cocktail party I was half-seriously defending the proposition that all undergraduate students should be required to take philosophy of science in order 'to learn methods of criticism and how to search for the truth.'

'Then they must certainly also take anthropology,' replied a colleague of that persuasion. 'We teach cultural relativism.'

I quickly admitted that cultural chauvinism was obnoxious and that we all needed to learn to approach other people with an attitude of respect and tolerance. I described a student comment on a course in Ancient and Medieval Science – 'I found out that Aristotle was no dummy' – and deftly drew the distinction between admiring the man and accepting his theory of falling bodies. And of course we had much to learn from other contemporary cultures, especially about caring for old people and integrating children into the workplace.

'No, I mean much more than that,' the anthropologist replied. 'I mean cultural *relativism*. All cultures are on a par, cognitively and morally. You can't judge one in terms of another.'

Anecdote 2: At an anthropology lecture, William Arens defended the controversial thesis that there is little direct evidence for cannibalism – much of the literature consists of travellers' tales, gossip about the cannibalistic practices of enemy tribes, and inconclusive anthropological observations of natives going off behind huts and preparing dishes which they *say* contain human flesh[1].

'But the latter remark destroys your whole argument,' remonstrated a member of the audience. 'If people *think* they are eating the brains

Currie, G. and Musgrave, A. (eds.), Popper and the Human Sciences.
 ISBN 90-247-2998-X

of their ancestors, then cannibalism exists.'

The speaker drew the obvious distinction.

'I'm not sure it makes sense to talk about what "really happened",' said the critic. 'In any case it is of no interest to social science. Our facts are entirely constituted by the shared ideas of the culture under study.'

The traditional philosophical theses of idealism and relativism have been severely criticized. Although it is impossible to refute such theories once and for all, it is possible to show that they have very little philosophical explanatory power. Yet anthropologists (who, like Aristotle, are no dummies, I presume) persist in paying a good deal of lip service to cultural relativism and cultural idealism. Why is this? This paper will present a partial solution.

My thesis is two-fold. I will first argue that *as a research strategy* cultural relativism is more successful than philosophers might like to admit. However, I will then go on to argue that the relativist programme is *not* enough for social science and that many anthropologists seriously over-estimate its heuristic power. Cultural idealism will be dealt with much more briefly and with much less sympathy.

2. *A brief summary of the literature.* I would not like to give the impression that anthropologists have not reflected upon and criticized cultural relativism. Often a distinction is drawn between toleration and moral relativism and it is frequently pointed out that even Franz Boas and Ruth Benedict, who originally defended a very strong version of cultural relativism, eventually fell into disarray because they wished to argue that there was something objectively bad about Nazi culture. Bagish claims that although the doctrine of moral relativism may have helped free us from 19th century ethnocentrism, today it is most often invoked in a reactionary fashion[2].

In an advanced sourcebook, Paul Schmidt gives a careful philosophical analysis of a variety of relativist theses and criticizes all of them except methodological relativism[3]. Kaplan and Manners in a theoretical monograph criticize what they call the ideological version of relativism and argue that 'anthropology must surmount the excesses of relativism'[4]. However, they also endorse relativism as a 'methological precept' and claim that 'all anthropologists accept the methodological version'[5].

What exactly is the principle of methodological relativism? Kaplan and Manners describe it as providing 'a useful reminder that in studying cultures other than our own we must try not to be swayed by our cultural preconceptions'[6]. Fair enough, but Schmidt says the principle prescribes that the anthropologist 'attempt to become a part of the culture under study, accept its values, traditions and beliefs in order to achieve a full "inside" understanding'[7].

If taken seriously, such a methodology falls prey to Popper's criticisms of the 'bucket theory of the mind'[8]. Marvin Harris, with characteristic forthrightness, labels as 'preposterous' the assumption that 'reliable descriptions of cannibalism and infanticide cannot be achieved by ethnographers who openly oppose these practices'[9].

It seems that what anthropologists call methodological relativism is little more acceptable than the ideological or normative thesis of relativism. What then of my promise to present at least a partial defence of cultural relativism? To do so, I must characterize what I will call cultural relativism as a *research strategy*. (This is also a methodological position, but to avoid confusion I will not use that term.) This is an explanatory programme which refers only to the native's beliefs, values, etc. Whether the scientist goes native in order to gain access to the culture is an entirely separate question.

3. *Motivation for the relativist research programme.* In order to explain an organism's behaviour there are a variety of factors which we may invoke. Here are some of them (this list is neither exhaustive nor exclusive):

a) Perhaps the behaviour is largely instinctual. If so, we will need to refer to the evolutionary history or genetic constitution of the species to which the organism belongs.

b) Perhaps the behaviour is largely the result of conditioning. If so, we will need to refer to the history of rewards and punishments which the individual animal has experienced.

c) Perhaps the behaviour is largely the result of pharmacological stimulants. If so, we will need to refer to chemicals in the individual's bloodstream, etc.

Each of the above modes of explanation may be employed with both humans and other animals. However, there is an additional mode of explanation which is frequently appropriate to understanding human behaviour, but only rarely (or minimally) useful in the case of animals. This is the case in which the organism's theory of the world (general conceptual system plus specific beliefs) influences its behaviour.

I will introduce this mode of explanation with an atypical example, a case where we might wish to say that something like a concept influences animal behaviour. Most dogs who have been housebroken receive their training in a very limited set of circumstances, e.g. one house with carpets and linoleum floors. Some dogs easily generalize the prohibition to include roofless patios, other people's houses, gazebos (with grass floor) and barns (sawdust floor). Others form a more limited 'concept' of what is forbidden and have to receive additional training covering automobiles, post offices and back porches.

To attribute concepts to dogs is somewhat controversial; however, in the case of humans we may ask them to report on the categories which are operating. I do not wish to suggest that our informants are always correct in their analyses and descriptions of their own behaviour. What I do maintain is that just as dogs with identical training behave as if they have different ideas of what a 'house' is, so may people in identical physical environments behave as if they have different ways of conceptualizing the world. And whereas we can not discover the various doggie 'concepts' of house by asking them, we may gain some information about the concepts influencing human behaviour by learning their language and questioning them. Furthermore in the case of humans, each individual does not start from scratch in forming a theory about the world. Most of our ideas are transmitted to us from our culture.

These truisms can be summarized as follows:

1) Much of an individual human being's behaviour is influenced by his/her theory of the world.

2) To understand such behaviour the scientist needs to know the relevant parts of that theory.

3) Much of an individual's theory of the world is influenced by his/her culture.

4) Therefore the social scientist's explanation will generally refer to, or relate the behaviour to, that culture.

I will call this the doctrine of explanatory cultural relationalism, i.e., explanations will often have to *relate* behaviour to culture. Behaviour depends in part on culture. An individual's behaviour generally must be understood *relative* to his/her culture. This I believe is the most plausible claim which can be disentangled from the vague slogan of cultural relativism.

4. *Cultural relativism as a research programme.* I now wish to examine a strong version of explanatory relationalism. It says that if we wish to explain the *typical* patterns of behaviour of individuals, *all* we need to

refer to in the *explanans* are other elements of that culture. The standards, beliefs, etc. of our own culture or modern science are strictly irrelevant – they do not appear within the explanation. Methodological relativism (which I defined and criticized above) denies that they even can be of heuristic value in discovering explanations.

Although such an explanatory programme has a rather repugnant idealist flavour to it, upon reflection it seems surprisingly plausible. Why are those students jumping in freezing water in mid-December with all their clothes on? Because (they believe) their soccer team has won the national championship and because (they believe that) jumping in Showalter fountain is the traditional way to celebrate (and they want to celebrate).

Although we often omit the phrases in parentheses in the explanatory sketch given above, note that it does not matter in the slightest whether these beliefs are true, well founded, etc. If the team has won but the students don't know it yet, they will not celebrate. If the report of the victory is mistaken or a cruel hoax, the students will celebrate anyway. As Popper made clear in his later expositions of the Rationality Principle, it is the *perceived* situation which explains, not the real one[10].

But what if we now ask for an explanation of the agent's beliefs and/or goals? If we are dealing with behaviour which is normal within the society under study the answer will surely refer to shared elements within that culture. Why do students want to celebrate sports victories? The answer might be given in terms of the values their society places on competition and team work. Or perhaps sports events have religious significance. Why do they believe they have won? Perhaps their society places great reliance on radio news broadcasts. Or perhaps they are the followers of a highly regarded sports guru who had a premonition that the team would win.

Again the intrinsic reasonableness of people's beliefs or goals seems to be irrelevant to the explanation of why they have them.

We seem to be sinking deeper and deeper into the idealist quagmire. Is there never a point at which we need to appeal to real truth, real beauty and real goodness? Are we doomed forever to tracing explanatory connections within a culture and never stepping outside of it?

5. *Universal elements within the cultural relativism programme*. The explanation sketches which I have given above skip over two places

where extra-cultural reality intrudes. First of all, the original *explanandum* says that the students in fact jumped into the water, not that they just thought they had. If we are to explain the fact that the students not only tried to jump in water, but actually succeeded, we obviously have to insert a premise about the real nature of Showalter fountain. At this point the die-hard relativist might object, saying that strictly speaking the *explanandum* event should be described as the agent would describe it, not in scientific or trans-cultural terms. Thus if Don Juan told Carlos Castañeda that he flew over a mountain, the anthropologist's job is to try to understand that flight, not to criticize his report or to explain why Don Juan thought he was flying. I will call this the doctrine of the culture-leadenness of *explanada*. Like the doctrine of theory-ladenness of observation which has plagued philosophers of physical science, there is a little bit of truth in it. However, just as holders of incommensurable paradigms in the physical sciences can find a level of description in which they can communicate (for example, one may have to speak of 'gas release' instead of 'oxygen production' or 'phlogiston absorption') so I believe people of different cultures can find an acceptable common level of description. Atheists have at least an acceptable minimal understanding of prayer and even grumpy pragmatists have *some* conception of out-of-body experiences. (Recall Bertrand Russell's remarks about sunsets and transcendental feelings.)

My main objection to the requirement that *explananda* must be expressed in native terms is that by conflating what Don Juan thinks he's doing with what he's really doing, we cut in half the number of explanatory questions which can be posed. In so doing we lose what to me are the more interesting ones, namely, why is Don Juan mistaken about what he is doing. More on this point in the next section.

There is a second place where extra-cultural reality intrudes into the cultural relativist explanatory programme and this is through the covering law or rationality principle which provides the major premise of the *explanans*. Philosophers have presented numerous variants of this explanatory principle ranging all the way from Popper's RP ('agents act more or less appropriately to their perceived situation')[11] to Churchland's L which has seven antecedent clauses[12]. Note it is not necessary that the agents *believe in* RP (or L) or consciously adopt it as a canon of rationality. All that the explanation requires is that the agent's behaviour conforms to it. (Thus the RP functions in explanations exactly like the law of falling bodies. Agents don't have to *know* about it in order to behave in accordance with it.)

It seems to me that the relativist might well accept this formal point but stress how insignificant it is for practical purposes. As Popper pointed out when discussing the Rationality Principle, all of our scientific effort goes into filling in the concrete details of the agent's situation. The Rationality Principle which 'animates' the model is taken for granted.

Our discussion so far may be summarized as follows: Although the strong thesis that *only* culture relative elements are required to explain typical native behaviour has been criticized, the trans-cultural elements needed, namely low-level descriptions of the behaviour and the Rationality Principle, are not very controversial or interesting. And if our explanations are criticized or improved upon it is rarely the trans-cultural elements which have to be modified.

6. *The limitations of the relativist programme.* In trying to make cultural relativism plausible as a research strategy I may have already changed its flavour to such an extent that many relativists would disown it. First, I dropped the *verstehen* methodology, saying that it was perfectly acceptable to use a variety of methods for ascertaining natives' beliefs and aims. Secondly, I pointed out that the major premise in the *explanans* was a universal law about how humans match means and ends and was independent of the native's culture. Thirdly, I permitted the scientist to describe the *explanandum* event in as objective a manner as possible. (This is not to deny that the scientist's attempt to give true descriptions is fallible.)

The resulting research strategy, although bearing a relativist label, bears a surprising similarity to many applications of Popper's method of situational logic. See, for example, Jarvie's complex situational analysis of cargo cults[13]. In short, one explains actions, and in some cases beliefs and values by relating them to the agent's subjective situation. As Jarvie puts it, he 'believes in beliefs'.

However, there is good reason to believe that sometimes agents can be as mistaken about their own reasons and motives as they are about the true nature of their actions. The most dramatic examples of cultural rationalizations have been provided by Marvin Harris[14]. For example, in certain Hindu villages where there is no need for draft oxen the mortality rate of male calves is almost twice as high as that of females. The villagers deny that they are practicing preferential male bovicide. Instead they claim that males die because they eat less. In areas where oxen are valuable, just the reverse explanation is proferred.

Whenever the agents themselves are mystified about why they are doing something the culture relativist programme will also fail to provide a correct explanation.

Except for cases such as those described above (how frequent are they?), the relativist programme seems to work as long as we stick to normal behaviour (which is explained in terms of widely accepted goals and beliefs) and low-level cultural beliefs/goals (which are explained in terms of deeper ones). In other words, the programme works pretty well as long as we stick to the problem of tracing explanatory connections within a complex cultural web. But what if we wish to describe and explain the origins of that web or changes within it?

Some social scientists would disavow both questions, saying there are not enough data to ask historical questions about origins and maintaining that it is almost a defining characteristic of *traditional* societies that they change very little or at a negligible rate. But even if these claims were true (I happen to disagree with both of them), we might still ask for deeper explanations or comparative accounts of the various static cultural webs.

7. *Alternative programmes.* There are a variety of research programmes in anthropology which attempt to go beyond cultural relativism – to ask questions which in principle relativism cannot answer. For example, sociobiologists look for shared biological propensities to avoid incest, insure paternity, rear healthy off-spring, etc. Cultural materialists, such as Marvin Harris, look at how societies solve basic or infrastructural problems, such as how to organize modes of production and reproduction, the assumption being that all human beings place a high value on basic necessities such as getting enough food (and the right kind of food – Harris believes both protein and fat needs have played an important role in the development of cultures).

Even Lévi-Straussian structuralists, who at times seem only to be interested in formal transformations of mythic elements, also talk about the importance of reciprocity in human societies, the implication being, that humans not only practice reciprocity, but also value it. And all functionalists postulate societal maintenance as at least a tacit goal of all cultures. Some, such as Parsons, would add other goals such as, value consensus, and integration to the list of universals.

Each of the above attempts to construct a general theory of culture has its weaknesses. Perhaps in reaction to the excesses of cultural relativism, sociobiologists and cultural materialists tend to underplay the causal role of beliefs in society. Their programme is to show how beliefs are determined (in part) by biology or problems of production and reproduction. They do not emphasize the fact that, say, religious beliefs can change biological or technological praxis.

The functionalist programme has been criticized on many levels – it does not account for change, power struggles, or the behaviour of minorities. The concept of latent function is intrinsically mysterious, especially when the hypothesized benefits are long-term or indirect. It also largely ignores the fact that people's ideas influence their individual behaviour and the institutions which they form.

Each of the above programmes promises to go beyond cultural relativism, but each ignores the basic strong point which cultural relativism insists upon, namely the fact that ideas influence human behaviour; they are *not* just epiphenomena.

8. *Towards a Popperian theory of culture.* Popper's philosophy provides a rich resource base for the student of social science. Like the culture relativist he emphasizes the importance of theories on behaviour. He argues that none of our ideas, even the low level observation claims, are independent of fallible theoretical expectations. And if even sophisticated, modern science is influenced by metaphysics why couldn't native botanical systems or theories of agriculture be influenced by magical or religious beliefs?

Popper has also stressed the importance of methodological norms and traditions on the development of ideas. Intellectual progress is not inevitable – it depends on societal structures which permit at least a modicum of criticism. But even with good scientific institutions, science is still dependent on luck – on there being people around who happen to come up with happy guesses.

What would a Popperian analysis of culture be like? First, we should look at the problems faced by a particular group of people. Some of these problems may be those of material existence. Some may be moral or religious problems. Others may have to do with removing unintended consequences of present institutions.

Then a Popperian might well look for the *constraints* which the culture places on solutions to these problems and its *methods* of evaluating solutions. Will only elders or priests be consulted? Are data to be collected or should one pray instead? There is, of course, no reason to believe that the same set of constraints and methods of appraisal will be used for every problem.

Given the people's problem situation, including constraints and methods of appraisal, we can generally understand why certain solutions and not others are adopted, although Popper also emphasizes the essential openness of the creative process. Problem solvers may come up with solutions that we would never have dreamed of; or they may miss solutions which appear quite obvious to us. Often, of course, the apparent lack of congruence between solution and problem situation is simply an indication that we, the scientists, have not correctly understood the institution or episode. (Natives are not dummies, either!)

The unpredictability of creativity poses one problem for the social scientist. There is another – one which Popper furnishes names for, but does not solve[15]. In discussing scientific discoveries Popper distinguishes between the scientist's *objective* problem situation and his *conscious* (or subjective) problem situation. For example, according to Popper, Kepler thought his problem was to discover the Pythagorean harmony of the world. The problem which he solved was that of providing laws of planetary motion.

What Popper does not make clear is which problem situation is to be utilized by the historian of science, or if both are to be used, how?

We can use Popperian language in describing Marvin Harris' example of the cows. The Hindus' objective problem situation is to not waste feed on unproductive animals. Their conscious problem situation is to raise lots of happy cows. But which problem situation best explains their behaviour? In this case, the objective one, but this is not always the case. How can we explain which takes precedence in a given case?

Once we allow objective problem situations to have causal import, why should we have scruples about explanations of institutions in terms of their latent functions, viz. positive effects which no one is conscious of, but which have an objective effect on the perpetuation of the institution?

As Agassi has so eloquently argued, the metaphysics of social science has too long been polarized into nature *vs*. convention, individual *vs*. the collective, rational *vs*. irrational[16].

Popper's method of situational analysis seems to offer a way of surpassing these dichotomies, for the logic of an agent's situation includes physical and biological factors, institutions, beliefs of all degrees of epistemological respectability, and goals, both idiosyncratic and communal. As such it is surely a valuable tool for social scientists. Yet the very richness of Popper's situational logic conceals a potential methodological hazard. Since we are permitted to refer either to the agent's objective problem situation, or his subjective one, or some combination of the two, it is imperative that we have independent evidence for our claim that the situational elements cited in our explanation are the ones which were actually causally efficacious. To return to the Harris case, it would be *ad hoc* merely to say, 'Material factors must be more important than religious ones in this case, because they best explain the Hindus' calf-rearing behaviour!'

Cultural relativists always bet that ideological elements will dominate behaviour. Cultural materialists believe that the infrastructure will generally prevail. Popperian situational analysis allows the social scientist a richer, more flexible approach. What one needs now is a theory about which sorts of factors are more likely to be of explanatory relevance in which situations.

Notes

1. Arens [1979].
2. Bagish [1968].
3. Schmidt [1968].
4. Kaplan and Manners [1972] p. 7.
5. *Ibid.* p. 6.
6. *Ibid.* p. 6.
7. Schmidt [1968] p. 174.
8. Popper [1972].
9. Harris [1968] p. 163.
10. Popper [1967].
11. *Ibid.*
12. Churchland [1970].
13. Jarvie [1964].
14. Harris [1979].
15. Popper [1972] p. 179.
16. Agassi [1977].

9. GOOD AND BAD ARGUMENTS AGAINST HISTORICISM

Peter Urbach

1. *Introduction*. From time to time historians and political scientists, perhaps noting some striking similarities in the development of various societies, or regularities within the evolution of one civilization, have formed the idea that these are the key to a universal feature which could provide the foundation for a general law of history. And some politicians have even seemed to act on the basis of such theories, often with disastrous consequences. For instance, Marx is supposed to have claimed that all human societies pass through feudal, capitalist and communist stages, in that order. Comte also had a three-stage theory of society and, more recently, Arnold Toynbee has developed a complicated hypothesis describing the various steps in the development and disintegration which all so-called 'civilizations' traverse.

One way of criticizing theories of this kind is to scrutinize each one separately to see whether it is testable and if so whether it has any inductive support. Indeed, Popper did mount a highly effective critical assessment of one version of Marxism, showing that it has been refuted[1]. Later manifestations of the theory were judged by Popper to be unfalsifiable, and hence unscientific[2]. Toynbee's theory too has been subjected to much criticism[3].

In this paper I shall not attempt to judge the merits of these objections. Instead, I wish to examine an alternative to this piecemeal kind of appraisal of historicist theories which Popper has explored in his famous book *The Poverty of Historicism*, according to which the whole approach to social science typified by these theories is wrong. If this comprehensive criticism were justified, then any hypothesis or prediction which fits the historicist pattern could be known to be unscientific from the outset without any further or more detailed examination of its particular claims.

Currie, G. and Musgrave, A. (eds.), Popper and the Human Sciences.

Popper's 'fundamental thesis' is that 'there can be no prediction of human history by scientific or any other rational means'. More particularly, Popper objects to what he calls 'the doctrine of historicism' which he characterizes as:

> an approach to the social sciences which assumes that *historical prediction* is their principal aim, and which assumes that this aim is attainable by discovering the 'rhythms' or the 'patterns', the 'laws' or the 'trends' that underlie the evolution of history (Popper [1960] p. 3).

That Popper should be the leading opponent of all historicist theories is perhaps surprising, for they seem capable of possessing all the virtues which he has praised in scientific theories. Consider for example a simple type of historicist hypothesis: 'all human social formations of type T pass through stages $S_1, S_2, ..., S_n$ in a given temporal order'. It is clear that this may be made more precise in ways which would turn it into a highly falsifiable and strictly universal generalization. So by Popper's standards of science, which he has discussed at length in other contexts, there can be, or so it would appear, nothing intrinsically unscientific about historicist theories. Nevertheless, Popper has advanced several reasons why he thinks historicism is wrong and these will be considered (and rejected) in what follows.

My own view is that Popper has been completely unsuccessful in his objective and that his falsificationist methodology of science not only lacks the resources to condemn historicist theories but actually establishes them as candidates for high scientific status. But, as I don't believe that historicist theories are, in general, scientific, I shall argue that this misappraisal exposes a serious shortcoming in the falsificationist philosophy of science. I will also show that a satisfactory explanation for the unscientific character of historicism is provided very naturally by a rival methodology to Popper's, namely probabilistic inductivism. In brief, then, my thesis is that Popper was right about historicism, but for the wrong reasons.

2. *The fallibility of predictions.* One extraordinarily poor argument which Popper has offered against the possibility of making scientific forecasts of historical events is based on the fact that they may be wrong. For instance, suppose an historicist notices similarities in the evolution of various societies and, as a result, proposes the hypothesis that all social groups of a given sort develop in the same specified way. Will this theory not form a legitimate basis for scientific predictions? No, according to Popper, for although 'history may sometimes repeat

itself in certain respects' and while 'the parallel between certain types of historical events, such as the rise of tyrannies in ancient Greece and in modern times, can be significant for the student of the sociology of political power' we should not forget that 'all these instances of repetition involve circumstances which are vastly dissimilar, and which may exert an important influence upon further developments'. Popper concludes that 'we have therefore *no valid reason* to expect of any apparent repetition of an historical development that it will continue to run parallel to its prototype'[4].

A similar kind of rebuttal is made against an historicist who confines his interest to a single society. For instance, a particular pattern or trend may have been observed over a long period of time and engendered the sociological conjecture that this pattern is a general feature of the society. But, Popper observes, 'a trend ... which has persisted for hundreds or even thousands of years *may* change within a decade, or even more rapidly than that'[5].

Both of these objections to historicist predictions rest on the accusation that the evidence upon which they are based does not validly imply, or guarantee, the truth of the predictions. In other words, the predictions might be wrong. But this fallibility is a characteristic which historicist predictions share with every prediction made in science. Indeed it is the very fallibility of our conjectures which lies at the heart of Popper's approach to science, so he can hardly turn it into a criticism of historicism[6].

3. *Historicism and physical necessity*. Popper employs another line of attack which takes issue with the pretensions of historicists to be able to find *physically necessary laws* of social evolution. So although Popper might be prepared to concede that an hypothesis about the stages through which societies pass could be corroborated and form the basis of successful predictions, he would object that it cannot be more than an accidentally true generalization.

Popper has given an example of a generalization which, if true, is only fortuitously so: 'all moas die before reaching the age of fifty'[7]. On the assumption that all the moas in the world are killed by an accidentally or contingently copresent virus, this universal statement is not a law of nature, but merely an accidental truth; it can lay no claim to describe necessary or structural features of the world. Had the fatal virus not happened to infest the habitats of all moas, they could have survived beyond the age of fifty. In contrast with the moa speculation,

Newton's mechanics and dynamics, if they are true, are laws of nature which remain true irrespective of the particular conditions which happen to prevail at any moment in history.

Popper's more precise formulation characterizes a law of nature as a general statement 'which is satisfied in all worlds that differ from our world, if at all, only with respect to initial conditions'[8]. Although Popper did not publish his theory of lawlikeness until 1959, well after the first exposition of his views on historicism, there is every reason to believe that this theory supplied the basis of his objection that historicist hypotheses cannot be laws. For instance, in keeping with his later characterization of physical necessity, Popper criticizes historicism on the ground that its 'laws of development' are not really laws at all but 'trends' extrapolated into the future in a way which

> overlook[s] the dependence of trends on initial conditions. They [historicists] operate with trends as if they were unconditional, like laws. ... This, we may say, is the central mistake of historicism. Its 'laws of development' ... do not depend on initial conditions (Popper [1960] p. 128; original italics removed).

What Popper presumably means is that although historicists ignore the fact, so-called 'laws of development' do depend on initial conditions and hence they are not really laws. If true, then they are merely accidental generalizations. This point is expressed more forcefully in the following argument:

> *The crucial point is this*: although we may assume that any actual succession of phenomena proceeds according to the laws of nature, it is important to realize that practically no sequence of, say, three or more causally connected concrete events proceeds according to any single law of nature. If the wind shakes a tree and Newton's apple falls to the ground, nobody will deny that these events can be described in terms of causal laws. But there is no single law, such as that of gravity, nor even a single definite set of laws, to describe the actual or concrete succession of causally connected events; apart from gravity, we should have to consider the laws explaining wind pressure; the jerking movements of the branch; the tension in the apple's stalk; the bruise suffered by the apple on impact; all of which is succeeded by chemical processes resulting from the bruise, etc. The idea that any concrete sequence or succession of events (apart from such examples as the movement of a pendulum or a solar system) can be described or explained by any one law, or by any one definite set of laws, is simply mistaken. There are neither laws of succession, nor laws of evolution (Popper [1960] p. 117; my italics replace those in the original).

Popper's meaning is, I think, roughly the following. Practically every evolutionary sequence in the physical or social world has its precise character determined by a variety of initial conditions; however, if different conditions had prevailed, then the evolution could have taken a completely different path. Thus the sequence of steps in the development are not nomologically linked, but just happen to follow one another.

I don't think that this argument bears close scrutiny, but nevertheless it seems that it might be the basis of a more interesting objection than the first one we considered. However, it cannot be sustained, for Popper's characterization of laws turns out to be quite unsatisfactory. As I mentioned, a universal theory is said to be a law if it is true in all worlds which differ, if at all, from ours with respect only to initial conditions. But what is it for a world to differ from ours *only* with respect to initial conditions? Clearly it is to say that that world differs from this one in its initial conditions but that it is governed by the same laws. Popper's definition is circular, then, in a way which vitiates its power to characterize laws: the definition of a law uses the notion of worlds differing in respect of initial conditions *only*, and this, in turn, depends on the idea of a law. Indeed, there is reason to think that any 'possible world' characterization of a law is bound to be circular; for if the law statement is not to be a logically necessary truth, there must be some worlds in which it is false. But we have no information to help to pick out the worlds in which the law holds from those in which it doesn't, except that the former is governed by that law[9].

But even if this difficulty could be circumvented, it is unlikely that it could be used in a criticism of historicism in the manner in which Popper has tried to use it. Pointing out that a theory is true only in this world, but false in certain others can hardly be a very powerful objection to scientists whose enquiries are directed to the goings-on of the actual world and who can have no access or contact with alternative existences. Moreover, in view of the complete causal isolation of possible worlds, there is no prospect of checking whether a theory is a law or not, nor of gathering any evidence relevant to the issue[10]. This of course applies to all theories, including historicist ones.

4. *A failed 'refutation' of historicism*. Popper's most influential argument and the one which claims for itself the most decisive effect against historicism appears in the Preface to *The Poverty of Historicism* and was invented only after the main text had already been published elsewhere. Unlike his other attempts to undermine historicism, this

argument is advertised as a 'refutation' since it supposedly demonstrates that 'for strictly logical reasons, it is impossible to predict the future course of history'. Let us consider this 'refutation'. It proceeds as follows:

(i) The course of human history is strongly influenced by the growth of human knowledge.
(ii) We cannot predict, by rational or scientific methods, the future growth of our scientific knowledge.
(iii) We cannot, therefore, predict [by rational or scientific methods] the future course of human history.

It is not difficult to see, however, that, as it stands, this argument is invalid. To be correct it would need some further assumption to the effect that events cannot be scientifically or rationally predicted when they are 'strongly influenced' by other events, like the growth of knowledge, whose occurrences themselves cannot be scientifically predicted. No doubt this is the kind of assumption Popper took for granted as uncontroversial and hardly worth stating when he put forward his 'refutation of historicism'. But it seems to me that this was a mistake; the assumption is both interesting and, more important, it turns out to be wrong[11].

To see this, consider what constitutes a rational or scientific prediction within Popper's theory of science. This is a prediction derived from a strictly universal, and falsifiable statement or one made highly probable by a general, statistical hypothesis. Although he is not completely explicit on this point, Popper may wish to add the requirement that for this statement to be the basis of scientific predictions it ought to be corroborated and unrefuted. Popper does not require the universal generalization upon which a scientific prediction is based to be true; were this kind of condition laid down, we could never know whether a prediction is scientific or not, and most of the celebrated predictions made by scientists in the past would presumably be unscientific.

If we put this construction on the idea of a rational or scientific prediction then Popper's objection to historicist predictions collapses. For if we consider any sequence of events then it is obvious that we can formulate a generalization about the members of that sequence and that this generalization may be both falsifiable and corroborated. Any prediction based on the generalization concerning later members of the sequence will therefore be scientific or rational. Moreover, even if the generalization happens to be false, any number of its predictions may be true. What is important here is that this holds whatever the nature of the events in the sequence, be they physical or social and whether or not their occurrence depends on unpredictable happenings. Each event may actually be determined by the toss of a fair coin or the whim of an angel or the radioactive decay of an atom of radium; nevertheless it can be scientifically predicted, in the sense in which Popper uses these terms.

Popper's argument, then, does not demonstrate the impossibility of making rational or scientific predictions of future social events. Even if it could be shown that these social events are 'strongly influenced' or even fully caused by unpredictable phenomena, this would not in the least undermine the possibility that these can be scientifically predicted, when interpreted within Popper's falsificationist philosophy.

5. *Probability arguments against historicism.* Although Popper's argument does not establish what he wants, we may nevertheless find his conclusion extremely plausible. To claim that every society of some rather general sort goes through the same particular stages, or that our civilization will have a certain specific development over a long period of time may seem incredible and absurd, especially if the social stages referred to are not ones just vaguely described in broad terms of physical transformations or energy changes but are rather specific social events such as the occurrence of wars, philosophical fashions, political institutions and the like. I think too that there is a good reason why such theories and the predictions based on them might be rejected, although this reason is not one which can be reconciled with Popper's methodological principles.

Consider, as a simple example, the scientific problem of predicting which team bats first in the England *vs* Australia Test cricket matches. A pattern, or trend, may have been detected, say that England was always the first team in whenever the game started on a sunny day. On the basis of this striking observation, we may be led to the 'bold conjecture' that this is a general rule and employ it to derive scientific predictions. These may even turn out to be true, thus corroborating our theory.

Now, someone better informed about the rules of the game will know that which team opens the batting in a cricket match is strongly influenced by the outcome of the toss of a fair coin. Let us assume for simplicity that the fall of the coin fully determines the batting order. In that case, the probability that any particular pattern appears is equal to

$(1/2)^n$, where n is the number of matches played. So we can see that the original theory linking the order of batting to the weather is, if true, purely fortuitous and in the light of the new information about the causal factors involved, the theory is very unlikely to be true – the actual value for the probability of the theory depending on n, the number of matches which are played.

This artificial example may perhaps illustrate the intention behind Popper's so-called 'refutation' of historicism. Many social developments are apparently dependent on chance-like phenomena, such as for example the creation of a particular scientific theory may be. If there are several stochastic processes determining a particular social event, or if there is just a single cause whose probability of occurring is very small, then that event may be unlikely to happen. We would then have excellent reasons for not relying on such a prediction and for not believing the generalizations upon which it is founded.

These reasons may be based upon either of the two main alternative interpretations of chance. The first of these looks on 'chance-like' phenomena as being irreducibly stochastic in virtue of an 'essential' indeterminism rooted in nature. In this case the probability of an event whose occurrence is dependent on them will be an objective fact about the world. On the other hand, the probability attached to an event is held by some simply to reflect our uncertainty or what we should regard as the appropriate level of uncertainty that the event will occur.

Let me summarize what I think can be rescued from Popper's arguments against historicism. If we assume that among the causes of those social events, in which historicists are interested, is a large number of more or less independently operating chance-like and unlikely phenomena, then the probability that social evolution will take any particular course is very small. Under these conditions the historicist hypothesis will be very improbable, and in this sense, it may be regarded as unscientific. Moreover, any prediction of some particular future occurrence is also unlikely to be true. We might then say that the forecast of an event which would happen only if several independent, chance-like and improbable causes come into play is unscientific, in the sense that it is very unlikely to be true. These considerations would, in addition, give us a very good reason for not relying on or acting on the basis of a general historicist claim or on any prediction drawn therefrom.

This, I think, provides a plausible and reasonable objection to at least some types of social prediction and it also seems to me to be the idea behind Popper's arguments which gives them their great plausibility. Nevertheless, strictly speaking, this objection is one which is not available to a falsificationist; for the extreme improbability of a theory is not a mark against it in this methodology, quite the contrary, the more improbable it is the better. And Popper has even attempted to demonstrate that all strictly universal scientific theories actually have zero probability[12].

On the other hand, falsificationism is not the only methodology which commands attention amongst philosophers of science. There are also consistent and well worked out methodologies according to which the probabilities of theories can and should be taken as crucial elements in their appraisal. Just such a methodology seems to have been applied by the economist Lionel Robbins in his objections to evolutionary or historicist theories in economics. His arguments in this connection are rather similar to Popper's, except that they are free from any falsificationist construction[13].

6. *Robbins on evolutionary theories in economics.* Robbins' view is that the economist can discover the types of causal factor which are relevant to certain economic phenomena, such as the rise in price or the surplus supply of a good, and that he is also in a position to find the laws by which these causes operate. Moreover, these 'economic laws' are strictly nomological or causal laws in the sense 'that they describe inevitable implications' and 'are on the same footing as other scientific laws'[14]. However, according to Robbins, laws are unlikely to be found to describe the changes which the causal factors will experience or to describe the path taken by a succession of phenomena each of whose occurrence is dependent on them. He gives an example to show why this is so. Suppose that the elasticity of demand for herrings in 1907-8 is found to be 1·3. The question arises: are we entitled to assume that this value will remain unchanged over a long period? Robbin's answer is that we are not, because the precise value for this elasticity of demand is

> a function of a great many *apparently independent* variables. It is a function of fashion ... [it] might be substantially changed by a change in the theological views of the economic subjects entering the market. It is a function of the availability of other foods ... of the quantity and quality of the population ... of the distribution of income within the community and of changes in the volume of money. Transport changes will alter the area of demand for herrings. Discoveries in the art of cooking may change their relative desirability. (Robbins [1935] p. 108; my italics).

Robbins also points to scientific inventions (for example, of the railway and of the internal combustion engine) and to alterations in our beliefs, volitions and feelings as independently acting influences on economic phenomena[15].

Moreover, these causal influences are themselves unpredictable:

> Economic Science knows no way of predicting out of the blue the configuration of the data [concerning these causal influences] at any particular point of time. It cannot predict changes of valuations [*i.e.* tastes and preferences.] (Robbins [1935] p. 122).

It is not that the nature of these influences is intrinsically unforeseeable – 'we need postulate no ultimate indeterminism' – it is merely that 'from the point of view of our system, such changes are unpredictable'[16].

In view of the many unpredictable and 'apparently independent' influences on the elasticity of demand for herrings, Robbins asks rhetorically:

> Is it possible *reasonably* to suppose that coefficients derived from the observation of a particular herring market at a particular time and place have any *permanent* significance – save as Economic History? (Robbins [1935] pp. 108–9; the first italics are mine).

Clearly Robbins thinks it is not reasonable to make such a supposition. The similarity to Popper's argument attempting to refute historicism is striking. However, unlike Popper he makes no claim here that we may not *conjecture* that the future will follow some particular pattern, nor does Robbins claim that we may not in fact make correct predictions using such a conjecture. His contention is that *we are unlikely to be right* either in the conjecture or in the prediction, in view of the many unpredictable and apparently independent causal factors operating to produce the phenomena in which we are interested.

Even if we should observe that some particular kinds of event have occurred together in several different societies at different times – for example the distribution chaos following price-fixing in relatively free markets which has been seen after both the French and Russian revolutions – 'this would not prove that the phenomena in question were causally connected in any intimate sense'. This point is essentially the same as the one of Popper's which I expounded in section 2, above. However, Robbins continues the argument: not only can we not validly prove a causal connection between phenomena which have been observed to coexist in different societies but, without further evidence, such observations do not even 'afford any *safe ground* for predictions with regard to their future relationship'[17].

The objection to such historicist theories then seems to lie in their very small likelihood of being true. The difference between them and genuine economic laws in regard to their scientific status is, according to Robbins, 'a matter only of *degree*'; the impression that there is a difference in kind being explained by the fact that the difference of degree is so great. Thus, although it '*might* be the case' that circumstances become sufficiently peculiar at some moment in history to make the supposed economic laws inapplicable, 'it is so overwhelmingly unlikely that we are warranted in neglecting the possibility'. On the other hand, Robbins continues,

> when we are dealing with the valuation [by members of an economic community] of particular products and the elasticity of demand derived therefrom, for the reasons already set forth, *there is surely an overwhelming probability that constancy is not to be expected* (Robbins [1935] p. 111; my italics).

Robbins adds that this holds *a fortiori* for more complex phenomena such as price fluctuations, cost dispersions and business cycles concerning which we are very unlikely to discover laws of development[18]. Concluding his analysis he writes:

> as regards our actual capacity to foretell a process of change, with its manifest dependence on the heterogeneous elements of contingency, persuasion, and blind force, if we are humble, we shall be modest in our pretensions (Robbins [1935] p. 135).

7. Conclusion. Comparing Popper's and Robbins' arguments, it is evident that they are based on roughly the same observations about the social world. Robbins' remarks are, to be sure, biased towards the economic sciences, but they are obviously more generally applicable[19].

Popper's employment of these observations to outlaw as unscientific hypothetical laws of social evolution is influenced by his falsificationist theory of science but, as I have already argued at length, this methodology is not up to that task. Robbins, on the other hand, does not discuss explicitly any general views about the nature and methods of science, but the standards he uses implicitly involve some version of probabilistic inductivism. For example, he frequently disparages economic theories of evolution for their improbability and contrasts them with economic laws which are 'overwhelming unlikely' to be false. Theories of historical development are also described as 'unreasonable' and as not having 'the same degree of probability' as economic laws, and predictions drawn from such theories are regarded as 'unsafe'.

The basis for these probability or plausibility judgements is the dependence of the phenomena comprehended by the theory on many apparently independent and unpredictable variables. The rationale for the judgements is, I presume, that each of the causes necessary to produce the particular economic event in question is unlikely to occur and since the numerous causes are 'apparently independent', the probability of that event is determined by the product of many probability terms, each of which is small. Hence the probability of the event itself is very small. I think that this is a correct account of Robbins' views; at any rate, it represents what I believe to be, and have already argued are, good grounds for mistrusting historicist theories.

This is not to say that historicist theories must remain forever unacceptable. It is possible for them to survive the severest tests and be confirmed repeatedly under a variety of circumstances, in which case the assumptions that the phenomena are determined by heterogeneous causes, or that these causes are independent will be undermined. Of course, these assumptions may also be rejected for other theoretical or empirical reasons, independent of any historicist theories. It is presumably because he has reason to think that there aren't manifold, independent causes of unemployment in the presence of 'some types of wage policy' that Robbins believes that 'it is quite certain' that unemployment results from such policies. He gives several other examples of circumstances where it is 'possible to predict with considerable confidence the actual results of given policies'[20]. Similarly, Popper instances the solar system as susceptible to long-term scientific predictions on the grounds that is is 'well-isolated' – that is to say, it is isolated from those unpredictable influences which are believed could affect the movements of the planet[21].

This is not the place for a general discussion of the relative merits of falsificationism and probabilistic theories of induction. However, the discussion brings to the fore a particular difficulty for falsificationism which is easily dealt with by its Bayesian rival[22]. For scientists would usually regard general theories of social or even physical evolution as 'absurd' and unworthy to be classed as scientific; nevertheless falsificationism makes the reverse valuation. On the other hand, as I have shown, probabilistic theories of induction do explain why and under what circumstances historicist hypotheses and predictions may be unsatisfactory[23].

Notes

1. See Popper [1966] volume 2.
2. Popper [1963] chapter 1.
3. See, especially, Pieter Geyl [1955].
4. Popper [1960] pp. 110 – 111; my italics replace those in the original.
5. *Ibid.* p. 115; my italics.
6. This point is made in my [1978a] section 1, and before this by Suchting in his [1972].
7. Popper [1959] p. 427.
8. *Ibid.* p. 433.
9. See especially Cohen [1980] for a clear exposition of this point.
10. This point is made with considerable force by Brian Ellis in his [1979] chapter III.
11. I have criticised Popper's so-called 'proofs' of his second premise, that we cannot predict the future growth of our scientific knowledge, in my earlier article where I showed that these arguments do not secure the desired result.

    Elsewhere, (Urbach [1978b]) I have presented arguments, in a more positive vein, which show that the role of random creative flashes of insight in the process of scientific invention is of much less significance than Popper and others have assumed: these philosophers overlook the availability of standard heuristic procedures in science. Elie Zahar has investigated, in considerable detail, heuristic techniques employed within relativity theory (Zahar [1980]).
12. Colin Howson has shown, however, that Popper's arguments to this end do not achieve their goal (Howson [1973]). Another reason for doubting the correctness of Popper's argument is that it depends on the assumption that scientific theories are always *strictly* universal and hence apply to an infinite number of instances. However the argument of Popper's which concludes that strict universality is required for a theory to have the highest scientific status is not a very satisfactory one. A preference for the more universal of two theories (assuming that neither has been refuted) 'is derived from the demand for the highest attainable degree of universality and precision, and it can be reduced to the demand or rule, that preference should be given to those theories which can be most severely tested' (Popper [1959] p. 123). But merely extending the scope of a theory, say to all regions of space and time, does not automatically increase its testability, except in the rather stretched sense that the theory could 'in principle' be tested by someone at any time or place. But an increase in the theory's falsifiability which permits an extension only in this 'in principle' kind of testing can be of no significance to us, and puts the theory in no greater peril of refutation. Moreover, if the greatest empirical content were a desirable goal for a theory, it would be all too easy to improve our scientific theories. For example, 'all planets and all Peruvians on Mars move in elliptical orbits' would be an improvement on Kepler's laws. This is clearly absurd.
13. Robbins [1935]. I am grateful to Paul Chauveau for drawing my attention to the relevance of Robbins' work.
14. *Ibid.* p. 121.
15. *Ibid.* p. 133.
16. *Ibid.* p. 134.
17. *Ibid.* pp. 73 – 4; my italics.
18. *Ibid.* p. 112.
19. Popper makes only a single, brief, reference to Robbins' work in the course of his criticisms of historicism, and this is to a brief article published in 1938. He does not refer to the much fuller statement of Robbins' ideas as expressed in his celebrated book, which was already in its second edition in 1935. Where he does mention Robbins, Popper shows support for his view that there are special difficulties in attempting to apply quantitative methods in the social sciences since the parameters whose values are required for prediction 'are themselves in the most important cases quickly changing variables' (Popper [1960] p. 143). Robbins' case, however, does *not* rest on the rapid alterations to which the values of these parameters are subject; his point is that we are not in a position to anticipate these changes.
20. Robbins [1935] p. 126.
21. Popper [1948] p. 340.

22. I have discussed some other facets of scientific method which falsificationism cannot explain but which can be accounted for within a Bayesian theory of science in my [1981].
23. The author is grateful to Colin Howson, Kurt Klappholz, Lionel Robbins, John Worrall and John Watkins for reading an earlier version of this paper and for making helpful comments which led to its improvement.

10. POPPER'S CRITIQUE OF MARX'S METHOD

W. A. Suchting

1. *Introduction: Popper's encounter with Marxism.* Marxism, according to Popper's own account, has played a quite special role in his life. In his autobiography[1] Popper recounts how, in the Vienna of the spring of 1919, he became convinced by communist propaganda and for 'about two or three months' regarded himself as a communist. But he further describes how some deaths during a pro-communist demonstration turned him against a view which claimed to be able to justify such deaths by allegedly scientific means. 'The whole experience ... led me to a life-long revulsion of feeling. ... By the time I was seventeen I had become an anti-Marxist.' Nevertheless, he goes on: 'The encounter with Marxism was one of the main events in my intellectual development. ... It made me a fallibilist, and impressed on me the value of intellectual modesty. And it made me most conscious of the difference between dogmatic and critical thinking.' Thus Marxism bears the credit not only for helping to make Popper intellectually modest, but also, in conjunction with more or less contemporary encounters with the work of Alfred Adler and Freud, for his formulation and answer to the 'demarcation-problem': what is distinctive about scientific statements is that they are open to critical assessment, in particular to empirical refutation (falsifiability principle).

Popper continued to criticize Marx and Marxism, notably in 'The Poverty of Historicism', written during 1935 – 1936, and then in a work he decided to write in 1938 but which did not see print till 1945, namely, *The Open Society and Its Enemies*, the second volume of which is of course devoted to Hegel and Marx. This in fact contains Popper's main critique of Marx, the earlier writing just referred to developing only one thread of it, and his other pieces on the subject being really no more than (to use an expression of a favoured author) 'parerga and paralipomena'. It is also the work which had the most public impact. Popper has referred to it as his 'war-effort'[2]. So indeed it was, provided we understand that the 'war' mentioned was not the one he meant but rather the Cold War. For in effect the book was a salvo on the intellectual front against what was seen as a threat from Communist Marxism both in the Soviet Union and Eastern Europe and also in the West (a front also occupied, in England, by writers like Isaiah Berlin, H.B. Acton and John Plamenatz with their weapons of various calibres). Though 'The Poverty of Historicism' was actually published slightly before *The Open Society*, it appeared in a small-circulation specialist journal and did not see book-form till 1957. By this time the larger work was already in a third edition, and, much more importantly, it was after the whole social and intellectual post-war scene had been shaken for good by the events of 1956: by the Suez invasion on the one hand and the revolts in Eastern Europe on the other (following upon the earlier reverberations of the revelations of the 20th Congress of the Soviet Communist Party). A new period of Marx-studies was now beginning as a result of the 'thaw' (however temporary) in both East and West.

The continuity between the work of this period and Popper's, to the extent that there is one, lies partly in Popper's general attitude to Marx. Thus he writes in *The Open Society* that:

> there can be no doubt of the humanitarian impulse of Marxism. ... Marx made an honest attempt to apply rational methods to the most urgent problems of social life. ... although he erred in his main doctrines, he did not try in vain.
> ... A return to pre-Marxian social science is inconceivable. All modern writers are indebted to Marx, even if they do not know it. ... even his mistaken theories are proof of ... his invincible humanitarianism and sense of justice. ... Marx's faith ... was fundamentally a faith in the open society (Popper [1966] volume 2, pp. 81, 82, 121, 200).

Nevertheless Popper believes at the same time that Marx was fundamentally mistaken.

It is at first sight surprising that Popper's critique of Marx still does not seem to have received much critical discussion – certainly not the sort that his accounts of those he regards as the two really deep-dyed enemies of the 'open society', Plato and Hegel, has. The volume in the Schilpp 'Library of Living Philosophers' devoted to Popper has only one piece on it, the which, by H.B. Acton, is predictably more than a little sympathetic. And the official Communist reply is really contained in an unremarkable book by Maurice Cornforth[3].

At any rate it is the aim of the present paper to make a contribution to this critical appraisal of Popper's critique of Marx as it appears in particular in the second volume of *The Open Society*[4]. This critique is to be found mostly in chapters 13 – 22 of that volume. It is laid out in three broad parts: 'Marx's Method', 'Marx's Prophecy', and 'Marx's Ethics'. In this paper I shall concentrate upon the subject-matter of the first part, which is both fundamental and reasonably self-contained.

2. *Popper's picture of Marx's 'Method'*. Although he does not say so in quite so many words, I do not think that we shall go far wrong if we assume that, for Popper, Marx's method, the method of historical materialism, is reducible to two main doctrines, namely (a) 'historicism' and (b) 'economism'[5].

(a) 'Historicism' is defined in one place as 'the claim that the realm of social sciences coincides with that of the historical or evolutionary method, and especially with historical prophecy'[6]. And earlier (marked in the index as a place where the meaning of a term is discussed) historicism is said to involve (and the context suggests that it is reducible to) the claim that '... it is ... the task of the social sciences to furnish us with long-term historical prophecies' based on laws of history[7]. So we may say, on the basis of these and other passages[8], that 'historicism' is sufficiently characterized by the following theses:

(1) The method of the social sciences is historical, a matter of the study of history.

(2) The study of history can vouchsafe knowledge of necessary, inexorable laws of social development.

(3) These laws form the basis of the 'prophecies', that is, the unconditional predictions about the future course of social development which it is the task of the social sciences to provide.

Popper also lists other views associated with historicism, though it is not clear to what extent these are definitory of that doctrine. One is a thesis regarding the *telos* of history which might well be listed as a further constituent of historicism, though it is not necessary to see it as part of its 'hard core':

(4) 'There is an inherent law of progress in historical development' and 'the development is one towards more freedom and reason'[9].

(b) 'Economism' is 'the claim that the economic organization of society, the organization of our exchange of matter with nature, is fundamental for all social institutions and especially for their historical development'[10]. (This sense of 'economism' is of course quite different from that usual in the Marxist tradition – see e.g. Lenin's *What Is To Be Done*? – but I shall use it in this paper in Popper's sense.) Popper tells us that

> ... owing to his Hegelian upbringing, [Marx] was influenced by the ancient distinction between 'reality' and 'appearance', and by the corresponding distinction between what is 'essential' and what is 'accidental'. His own improvement upon Hegel (and Kant) he was inclined to see in the identification of 'reality' with the material world (including man's metabolism), and of 'appearance' with the world of thoughts or ideas. Thus all thoughts and ideas would have to be explained by reducing them to the underlying essential reality, i.e. to economic conditions (Popper [1966] volume 2, p. 107 and cf. especially p. 326, n. 13).

Thus economism supplements the theses of historicism by specifying that the laws of development alluded to (in particular in thesis (2)) are laws of *economic* development. Consciousness is epiphenomenal upon or even reducible to the really real, which is matter, and in particular matter involved in production. A consequence is that since 'all politics, all legal and political institutions as well as all political struggles' are just a 'way in which the economic or material reality and the relations between the classes which correspond to it make their appearance in the world of ideologies and ideas', they 'can never be of primary importance. *Politics are impotent*'[11].

3. *On the likeness of Popper's picture*. An unavoidable preliminary question raised by all this obviously is: what licence is there in Marx's own texts to ascribe to him the doctrines of 'historicism' and 'economism'?

A first answer is: there are indeed passages – and ones from central texts – which may be cited in support of the view that Marx did subscribe to these doctrines. A sufficient number of such are well known. Thus, as regards historicism, in the 'Afterword' to the second German edition of the first volume of *Capital* he cites with approval a characterization of his method by a reviewer of the first edition: 'Marx considers the social movement as a process of natural history, guided by laws which are not only independent of people's will, consciousness and purposes, but which rather, on the contrary, determine their will, consciousness and purposes'[12]. And towards the end of the work in the great purple passage in the chapter on 'The Historical Tendency of Capitalist Accumulation' he writes how 'capitalist production begets, with the inexorability of a natural process, its own negation'[13]. Then, on the theme of economism, in the famous passage on historical materialism, in the preface to the *Contribution to the Critique of Political Economy*, Marx begins by distinguishing between, on the one hand, 'the economic structure of society', and, on the other, 'a legal and political superstructure' which arises on that 'real foundation' and 'definite forms of social consciousness' that 'correspond' to it. He goes on to speak of how the first changes from time to time and remarks: 'With the change of the economic foundation the entire immense superstructure is more or less rapidly transformed'[14], which at least strongly suggests the secondariness of the 'superstructure' with respect to the 'base'. Let these passages stand for others.

So a first answer to the question posed to start with is that there is indeed some textual support for the presentation of Marx as an historicist and proponent of economism.

But a second answer to the same question is: it is also very easy to cite other passages, likewise from central texts, which support a presentation of Marx as holding views opposite to these. Thus, as regards historicism, Marx wrote ten years after the initial publication of the first volume of *Capital* that the latter 'does not pretend to do more than trace the path by which, in Western Europe, the capitalist economic order emerged from the womb of the feudal economic order', and, moreover, that this 'historical sketch of the genesis of capitalism in Western Europe' must not be transformed into 'a theory, belonging to the philosophy of history, concerning the general path of development that is fatalistically prescribed for every people, whatever the historical circumstances in which it finds itself'[15]. And as regards economism we can turn, with respect to the political dimension, to the *Manifesto of the Communist Party* (the last edition of which by both Marx and Engels was the Russian one and appeared just one year before Marx's death, in 1882). Here, in the programmatic part, it is said very explicitly that 'the *first* step in the revolution by the working class is to raise the proletariat to the position of *ruling class*. ... The proletariat will use its *political* supremacy to wrest ... all capital from the bourgeoisie'[16]. That is, political change is envisaged as *preceding* change in the economic sector. Again, as regards 'forms of consciousness', ideology, Marx writes of the working class:

> The recognition of products as its own, and the judgement that its separation from the conditions of its realization is improper, forcibly imposed – is an enormous [advance in] consciousness, ... and as much

> the knell to its [capitalism's] doom as with the slave's consciousness that he *cannot be the property* of another, with his consciousness of himself as a person, the existence of slavery becomes a merely artificial, vegetative one, and ceases to be able to continue as the basis of production.[17]

Again let these passages stand for all of similar tenor.

4. *The theoretical heterogeneity of Marx's texts*. Supposing that the sorts of passages just cited are to be taken at their face-value, then the first and most obvious conclusion to be drawn is simply that Marx's texts are not theoretically homogeneous, that, to some degree at least, they contain different and even conflicting positions. This seems clearly to be the case with regard to different works; we may even suspect that it is at least sometimes the case with regard to one and the same work. If this conclusion is correct, as I think it is, then the next question is: what are the various theoretical strands from which Marx's texts are woven?

I suggest that there are at least three such strands and, insofar as we are interested in only the most important ones, perhaps at most three. There is no question of characterizing these in detail here but perhaps the following highly summary indications will suffice for the present purposes.

(*a*) *'Essentialism'*

The first occurs in Marx's work, to start with in Hegelian-idealist form. Thus in the notes to his doctoral dissertation he writes that 'the *practice* of philosophy is ... *critique* that measures the individual existence by the essence, the particular reality by the Idea'[18]. The same notion occurs in his early newspaper work. We must, he writes here, 'take the essence of the inner idea as the measure to evaluate the existence of things'[19]. In the *Economic and Philosophical Manuscripts*, written a couple of years later, Marx has become a materialist (of sorts) but the essentialism is still there even if it is now what may be called naturalistic rather than idealist: a theory of the essential nature of human beings, their '*Gattungswesen*', functions as a criterion for evaluating existing institutions. A heavily idealist-tinctured version appears again in the *Grundrisse* in, for example, Marx's discussion of the accumulation of capital[20]. (Indeed Marx himself jots down the remark at one point: 'It will be necessary later ... to correct the idealist manner of the presentation, which makes it seem as if it were a matter of conceptual determinations and of the dialectic of these concepts'[21].

But the idealism is not merely in the 'manner of the presentation' and cannot be got rid of as easily as this.) More generally, this essentialism is at least one of the sources of whatever tendencies there are towards 'historicism' and 'economism' in Marx's thought. For the essence may be conceived to determine the character of its self-realization, to constitute the originating centre of phenomena which are hence, in a certain sense, reducible to it.

(*b*) *'Structuralism'*

What I am calling here, for want of a better name, but perhaps still rather infelicitously, the 'structuralist' element in Marx is a matter of the primacy accorded to social *relations*. It finds compendious expression in a remark like: 'Society does not consist of individuals, but expresses the sum of relations (*Beziehungen*), connections (*Verhältnisse*) within which these individuals stand'[22], the individuals being regarded, in a complementary gloss[23], as the 'bearers (*Träger*)' of these relations. If 'essences' are thought of as pertaining to subjects rather than to relations between subjects, then 'structuralism' is a strand in Marx's thought quite different from essentialism. But if not, and structures are conceived of as being able to have their own essential characteristics, then structuralism can be regarded as a form of essentialism: structures rather than individual human beings or humanity may be conceived to have their essential characteristics that unfold inexorably and generate other structures which are in some sense reducible to them.

(*c*) '*Practical materialism*'

A third strand may be called 'practical materialism'. It is the subject matter of the notes from Marx's pen in spring 1845 which Engels described, in publishing them for the first time, as 'Theses on Feuerbach', the 'first document in which is deposited the brilliant germs of the new world-outlook'[24]. Here Marx writes that 'the chief defect of all previous materialism' is that the world is not conceived from the point of view of '*human activity, practice*'. Specifically: 'All social life is essentially *practical*. All mysteries which lead theory to mysticism find their rational solution in human practice and in the comprehension of this practice'[25]. That is, the crucial concept for the understanding of the social world is neither individuals of whatever sort characterized in essentialist terms, nor relations, but the material transformative activity, practice, of human beings[26]. This very naturally connects with the view of the centrality of the idea of class-conflict in understanding society and history, a centrality which is affirmed in the plangent opening of the *Manifesto of the Communist Party*: 'The history of all hitherto existing society is the history of class struggles'[27].

I have called these 'strands' and have used the metaphor of Marx's texts' being woven from them. This implies that they are not neatly separated as between texts or even within the same text. It also implies that one or other of them may be predominant in a particular text or over all or most of the texts of a certain period[28]. That there should be these theoretical fissures – the works of none of the master-thinkers are free of them – should be surprising only to someone who insists on trying to explain texts in terms of the fictitious unity of an authorial origin, whereby the different productions of a writer are conceived as expressions of a single essence which can, in principle, no more emanate inconsistent views than an oak-tree seed can sprout cabbages as well. If we do not hold such a view then a thinker's ß *euvre* becomes a complex product of many different generative conditions bound together into whatever unity it has by numerous factors of an individual-biographical and social sort (to whatever extent the two can be separated or even closely distinguished). Of course, an attempt might be made to preserve a certain essentialism of reading by holding that one or other of the strands represents the author's 'essential' ideas (destined for celebration or reprobation). But what are the criteria for deciding which of the ideas are thus privileged? I cannot see any viable answer to this question.

5. *What is to be done?* Are we left then between the devil of an essentialist reading and the deep blue sea of a merely descriptive listing of various components of a text? I do not think so. We can in fact go on to ask at least two broad sorts of questions. One is: which of the strands is or are most adequate for the solution of such and such problems, theoretical or practical? A second is: what were the conditions for the generation of these different doctrines in the work of a single thinker? and: what were the conditions for these different doctrines having had, each in its specificity, the intellectual and social influence it has (or failed to have)?

With particular reference to Marx, and regarding the first of the two questions raised above, I want to suggest that it is the third strand that is immeasurably the most adequate of the three with regard to the theoretical task of understanding socio-historical affairs and the practical one of changing them in certain directions. In particular I suggest that whilst the first two strands are to some degree compromised by historicist and economistic views with which they have a certain relation, the third is not. It is a basic principle of Popper's own methodological canon that a theory should always be criticized in its strongest form[29]. So I shall spend most of the rest of the paper sketching some of the main lines of what I take this third thread in Marx's thought to involve. There is naturally no question here of doing more than this[30]. But perhaps even a very rough sketch will suffice to give some plausibility to the above claims that it is not historicist or economistic and has more initial attractiveness than either of the other two.

6. *An outline of 'practical materialism'*. I shall try to spell out what I have called 'practical materialism' in the form of a few indicative theses with just sufficient reference to Marx's texts to make it clear that it is not a matter of merely ascribing to Marx a figment of my own imagination.

(a) 'All social life is essentially *practical*'[31]. That is, it is a complex of *practices*. The paradigm of a practice in Marx's work is of course *economic* practice (though it is not the only one). As this is explained in *Capital*[32], it has, in the first place, a structure the elements of which are (1) purposeful activity or labour, performed upon (2) a certain sort of object by means of (3) certain instruments of labour, in order to realize the goal of the process. The practice is, in the second place, a set of relations between these elements the character of which is determined by (i) the technical exigencies of the process and (ii) the type of social control exercised over the elements of the process ('technical' and 'social' 'relations of production'). Economic practice is but one of many practices or ways of transforming the world in accordance with human purposes by various sorts of instrumentation; in short, it is but one of many ways of 'appropriating the world'[33]. Thus political practice changes the world of social relations by means of, e.g., decision-making procedures, artistic practice transforms, for instance, certain patterns of sound into others with a view to changing human patterns of thought and emotion in certain ways. Ideological practices transform the manner in which people experience their relation to the real world. And so on.

(b) It was said in (a) that there are different *sorts* of practices. It would be more accurate to say that any practice falls under different descriptions of it *as* a practice, and the various aspects are distinguishable but not separable. Thus one and the same practice which may be described as that of producing cars on an assembly line is also a process of exploitation of labour-power[34]; at the same time it is a political practice from the point of view of the significance of the process in the general struggle between capital and labour for control of the situation; and it is an ideological practice from the point of view of its significance for reproducing or changing the dispositions of the people engaged. Insofar as the major significance of it (at a time) is the production of cars, it may be called an 'economic practice' *tout court*. Similarly, there are predominantly political or ideological practices which are also properly described in each of the other ways, thus 'economic', 'political', 'ideological', etc. in this context are *functional* descriptions[35].

(c) Within a practice it is necessary to distinguish between, on the one hand, the modes of representation of the practice by its executants, and, on the other, the objective character of the practice itself. These two are such that (1) the first may not be a veridical picture of the second, and (2) in any case the first is in general brought about or at least has its main features determined by the practice of which it is the representation and/or other practices. It is not that there could be a practice without some representations by the executants about what is going on, but rather that they need not have accurate – even approximately adequate – ideas about what is going on. (This is familiar from psychoanalysis.) 'They do not know it, but they do it' as Marx sums it up in his analysis of the structure of the exchange-relation[36]. This thesis of the primacy of the objective structure of the practice over representations of it is a fundamental sense of 'materialism' in 'historical materialism': the materialist 'conception of history ... does not explain practice from the idea but explains the formation of ideas from material practice'[37].

(d) A second sense of 'materialism' in 'historical materialism' is that these different practices form, amongst themselves and without reference to their internal structures, a system whose character is ultimately determined by the economic. This is the more familiar thesis: 'The mode of production of material life determines the social, political and intellectual life process in general'[38].

(e) Among the various objective features of practices are certain 'tendencies'[39] for these to change in certain ways. These are objective in the sense that their occurrence does not depend on the will of executants of the practice. (A simple example: to the extent that a market is non-monopolistic and the demand for a commodity to all intents and purposes static, there is a tendency for the increase in supply of that commodity to lower its price, independently of what the supplier might wish.)

(f) Whether a certain tendency manifests itself (in particular as what might be called a 'trend') depends upon the interaction of the effects of various practices[40]. These depend in turn on a large number of circumstances not exhaustively enumerable or determined by the character of any one practice. It is not determined in particular by the economic, which only puts limits upon the changes which can be introduced by other practices. In particular, in the case of the period of history with which Marx was mainly concerned, namely, the period of class-societies, the realization of tendencies is a matter of the outcome of the *class-struggle*, which certainly has economic *determinants*, without being exhaustively *determined*, as regards its outcome, by the economic. The possible efficacy of political practice is constrained by the economic but nevertheless has a degree of causal autonomy.

(g) A briefly sketched example may help eludicate the above all too compendious theses.

Take the case of capital-accumulation, which is, according to Marx, one of the two 'characteristic features' of the capitalist mode of production[41]. How is it rooted in the capitalist mode of production? What are the conditions for the possibility of its being not merely a tendency but a 'trend' also? In fact to answer the first is already to give a general answer to the second.

The feature that is constitutive of the capitalist mode of production is its being a generalized commodity-producing economy[42]. This implies in particular the occurrence of labour-power as a commodity, and that in turn the existence of a buyer and user of it, namely, capital. So the key relation is that between capital and wage-labour. Therefore, the most fundamental explanation of the tendency towards the accumulation of capital must lie in this relation. Accumulation of capital affords a basic means for the maintenance and expansion of the control of capital over labour through the increase it makes possible in the ratio between the amount of capital invested in means of production and the amount in labour-power. The former makes possible, in the first place, the transition from essentially hand-tool production to machinery-based production proper. This works in the direction of dispossessing wage-labour of the means of production not merely legally but also technically (the strength and skill of the worker becoming increasingly irrelevant to the production-process) and also broadens the base of exploitable labour-power (women, children, etc.). It also tends to counteract the tendency towards a rise in wages which would otherwise operate (because of an increase in the demand for labour-power) were an increasing scale of production to be carried on with the same relative amounts of capital spent on labour-power and means of production. Furthermore, the availability of increasingly large amounts of accumulated capital permits capital to wage the economic class struggle more effectively (capacity to ride out strikes, close plant permanently if necessary, etc.) and also to dampen political class-struggle by controlled increases in the real standard of living and thus – on the plane of ideological class-struggle – to legitimate the existence of capitalism. (There is much more to be said on this score, but this should indicate the general direction of explanation[43].) All this relates to the sphere of production. Within the sphere of distribution the increase in productivity through improvements in technology and increases in scale of production made possible by investment of accumulated funds makes possible a cheapening of unit commodities and hence better ability of individual capitals to compete with other capitals for realization of surplus.

Now the point is that all of these factors depend on the complicated interaction of struggles: most fundamentally, between wage-labour and capital, but also between segments of capital. It is the first that underlies the need for capital to find every possible means to maintain and expand its control over wage-labour. It is the state of the political and ideological practice of the working-class at different times and places that (partly) determines whether capital is able to accumulate surplus rather than yielding up a greater or less part of it to the working-class in the form of wages and/or better conditions; the state of that practice also determines whether labour sets itself goals other than improvement of wages and conditions within the capitalist system and looks towards the destruction of capitalism altogether. Even the level of accumulation made necessary by the exigencies of competition depends, amongst other things, on the degree of monopolisation in the economy, and this is partly but importantly a consequence of political and ideological struggles against such trends. (Again there is much more that could be said here but the above gestures towards the general lines of such further developments.)

(h) A society in which the capitalist mode of production prevails tends, at some point, to run into fundamental problems affecting its very functioning. The basic economic dimension of these problems consists in a conflict between, on the one hand, the tendency of capitalism to develop the social productive powers without limit (see the preceding point), and, on the other, the possibilities for utilising these powers, that is, the conditions for realising the tendency for development, both tendencies being rooted in the sort of control over the productive resources characteristic of capitalism. From the point of view of the non-owners of the productive powers, the only viable long-term solution to the problem is to take control of these powers from those who currently use them for the production of profit and to use them in a collectively controlled way for the satisfaction of their needs. This, considered from an economic point of view, would be to reorganise society along socialist lines. The present controllers of the productive resources naturally resist this sort of solution to the problem and try to carry through other solutions (lowering real standard of living, unemployment, war, and so on). This clash of real tendencies is expressed in class-struggle, the outcome of which at a time is unforeseeable and could involve, not the victory of one of the two sides, or some other, but 'the common ruin of the contending classes'[44].

7. *Popper's criticism of Marx's method and 'practical materialism'.* I claimed earlier than an historical materialism firmly rooted in Marx's notion of practice is not historicist and not economistic and therefore not subject to any of the objections to which they are. This claim can now be briefly vindicated in the light of the outline of that version of historical materialism in the preceding section.

Let us start with *historicism*. The practice-based version of historical materialism certainly holds the method of social science to be historical (in a sufficiently broad sense of the word), but it does not contain the thesis of the existence of unconditional, inexorable laws of social development. As we have seen, it certainly includes the idea of objective tendencies, but whether and how these are manifested is a result of definite human practices, the ultimate outcome of which is not predictable. As a result of this the task of social theory is not the annunciation of prophecies but rather the acquisition of an understanding of tendencies and conjunctures which – insofar as the social theory is carried on within the context of political practice – can be used to intervene in the course of history with a view to influencing it in one direction rather than another[45]. A Marxism of the sort sketched above has absolutely no use for long-term predictions – the very idea of which is indeed inconsistent with it – but only for (contextually) long-term aims. Finally, for such a view there is no overall teleology in history, no pre-destined End, but only innumerable dispersed, contextual ends which people set up in the course of solving their problems and the results of the pursuit of which interact to form the conjuncture which is just history at a certain time.

With regard now to *economism,* we have seen that on the above reading of Marx there is not even an actual topographically separate region of the economic to which other regions (politics, ideology, etc.) could be reducible or with respect to which they could be epiphenomenal. The non-economic sectors are no more and no less 'material' than the economic (considered as an aspect of material practices)[46] and the non-reducibility and non-epiphenomenality of the non-economic is guaranteed by the co-determination of each by all. In particular, to hold that the political is not of *primary* importance is by no means to say that it is 'impotent': to hold that certain political changes (at least of a relatively permanent sort) are not possible without certain economic ones is perfectly consistent with holding that certain economic changes are not possible without certain political ones (even preceding political ones).

8. *Conclusion*. In Section 5 above I posed two questions, for the first of which I have sketched the outlines of a reply. The second question concerns the socio-historical conditions for the existence (and co-existence) of the various components in Marx's thought; an extension of this concerns the conditions for the later historical influence of Marx's ideas or some interpretations of them. This is subject-matter for another inquiry. But a few quite tentative remarks about whatever tendencies there are to historicism and economism in Marx's thought may gesture towards the direction of further reflections about this problem. ('Whatever tendencies there are': the explicit expression of these is largely confined to some purple passages.)

As regards historicism I think that the right direction in which to look is that of the state of the class-struggle in Europe in the period after the defeat of the first major workers' struggles in the nineteenth century, namely, those which took place during 1848–49. The foundation of historical materialism as a practice-orientated outlook on society falls in the period of the build-up to 1848. Tendencies to historicism are mainly visible in the period after 1850,[47] with Marx's translation to London and an almost complete cessation of significant political activity on his part (reflecting his milieu) for nearly fifteen years (till the foundation of the International Working Men's Association in 1864, by which time his major economic work was completed in essentials). Such tendencies have a significant ideological function. As Gramsci wrote later, 'when you don't have the initiative in the struggle and the struggle itself comes eventually to be identified with a series of defeats, mechanical determinism becomes a tremendous force of moral resistance, of cohesion and of patient and obstinate perseverance'[48]. (It then becomes clear how very pertinent was the title of Gramsci's now famous article greeting the October Revolution: 'The Revolt Against *Capital*'[49].)

As regards economism the clue here is a combination of both the situation which led to historicism (what could the inexorable trends be trends *of* but the economic?) and of the theoretical context within which historical materialism was born. In a situation in which idealist theories of society and history were predominant an economism has a polemical import as the furthest possible 'bending of the stick' in the opposite direction. This is indeed the drift of some of the late Engels's reflections. 'According to the materialist conception of history, the *ultimately (in letzter Instanz)* determining element in history is the production and reproduction of real life. More than this neither Marx nor I have ever asserted.' He goes on a little later:

> Marx and I are ourselves partly to blame for the fact that the younger people sometimes lay more stress on the economic side than is due to it. We had to emphasize the main principle vis-a-vis our adversaries, who denied it, and we had not always the time, the place or the opportunity to give their due to the other elements involved in the interaction (Engels to J. Bloch, September 21 – 22, 1890; Marx-Engels [1956] 37: 462, 465).

But all this shows only that – as historical materialism itself holds – the process of generation of theory takes place in and through definite material – in particular ideological – conditions. After all, Popper's own doctrine of truth is explicitly presented in the context of its ideological role in the struggle against irrationalism and moral relativism[50].

Notes

1. Schilpp [1974] pp. 24ff.
2. Popper [1966] volume 2, p. 393
3. Cornforth [1968].
4. In an earlier paper entitled 'Marx, Popper and "Historicism" ', I directed criticism mainly against *The Poverty of Historicism*.

5. Cf. Popper [1966] volume 2, p. 106.
6. *Ibid.* volume 2, p. 106.
7. *Ibid.* volume 1, p. 3.
8. *Ibid*, volume 2, pp. 82f, 86, 136, 319, etc.
9. *Ibid.* volume 2, p. 319.
10. *Ibid.* volume 2, p. 106.
11. *Ibid.* volume 2, pp. 118, 119.
12. Marx [1976–81] volume 1, p. 101 (Marx-Engels [1956] 23:26). Cf. also *ibid.* p. 91 (Marx-Engels [1956] 23:12).
13. *Ibid.* p. 929 (Marx-Engels [1956] 23:791).
14. Marx [1970] p. 21 (Marx-Engels [1956] 13:8f).
15. Letter to *Otechestvenniye Zapiski*, November 1877 (Marx-Engels [1956] 19:111. Included in Marx-Engels [1953].) See also Marx's letter to Vera Zasulich 8/3/1881 and the three preliminary drafts, Marx-Engels [1956] 19:242f, 384ff.
16. Marx-Engels [1975] 6:504 (emphasis supplied), (Marx-Engels [1956] 4:481). See also countless passages especially in Marx's correspondence, for example, his letter to F. Bolte, November 23, 1871 (Marx-Engels [1956] 33:332f included in Marx-Engels [1953]). One of the basic lines of Marx's work in the International Working Men's Association was precisely the need for an independent workers' political party.
17. Marx [1973] p. 463; Marx [1953] p. 366f ('knell to its doom' is in English in the original). Cf. also Marx's letter to L. Kugelmann, July 11, 1868 (Marx-Engels [1956] 32:553f) (included in Marx-Engels [1953]).
18. Marx-Engels [1975] 1:85 (Marx-Engels [1956] Ergänzungsband I. Teil, 327f).
19. Marx-Engels [1975] 1:154 (Marx-Engels [1956] 1:50).
20. On the Hegelianism of the *Grundrisse*, see the paper on the latter (in the form of a review of Rosdolsky's *The Making of Marx's 'Capital'*) by John Mepham [1979].
21. Marx [1973] p. 151 (Marx [1953] p. 69).
22. Marx [1973] p. 265 (Marx [1953] p. 176). This recalls the sixth of the 'Theses on Feuerbach': 'The human essence is no abstractum dwelling in each single individual. In its reality it is the ensemble of social relations (*Verhältnisse*).' (Marx-Engels [1975] 5:4; Marx-Engels [1956] 3:6). But I am inclined to think that '*Verhältnisse*' has here its vaguer sense of something like 'circumstances'. (Cf. '*Umstände*' in the third thesis.)
23. Marx [1976–81] volume 3, p. 957 (Marx-Engels [1956] 25:826f).
24. Marx-Engels [1970] 3:336 (Marx-Engels [1956] 21:264).
25. Theses 1 and 8, Marx-Engels [1975] 5:3,5 (Marx-Engels [1956] 3:5,7).
26. In a not unfamiliar passage from the 'Notes on Adolph Wagner' Marx writes that his '*analytic* method ... does not start out from *man*, but from the economically-given social period' (Marx [1975] p. 201 = Marx-Engels [1956] 19:371). If we identify the 'economically-given social period' with the economic base of the preface to the *Contribution to a Critique of Political Economy* then we have here a nice expression of the contrast of the first two positions.
27. Marx-Engels [1975] 6:482 (Marx-Engels [1956] 4:462).
28. Any attempt to come to terms with the heterogeneity of Marx's work must also come to terms with the periodization introduced by Althusser in the early 60's and held to with only minor modifications ever since. In my opinion his work here (as in most places elsewhere) marks an achievement as a whole without parallel in the history of Marxist philosophy. Nevertheless, I differ from him here in at least the following respects. (a) Despite a few disclamatory murmurs here and there, he in general identifies theoretical differences with chronological ones in Marx's work. I want to distinguish the two more radically and see the texts themselves as more complicated 'mixes' of different views. Certainly, though, various works are more predominantly in one direction rather than another. (b) He fails to distinguish what I have called the 'structuralist' thread – probably largely because he has strong inclinations in this direction himself, though he emphasized in a seminal way the centrality of practice (even if mainly in its epistemological significance). (c) He has a tendency to look at Marx's texts in an essentialist way, rather than seeing their theoretical complexity in a fully historical fashion.
29. See Popper [1968b] p. 88: 'The most valuable kind of criticism is one that takes a theory, formulates it as clearly and sharply as possible, putting it in its best possible shape, as it

were, and which then proceeds to show that, nevertheless, there is something wrong with it. The least valuable kind of criticism is one that misunderstands or misinterprets a theory, and shows that there is something wrong with the theory when thus misunderstood or misinterpreted.'

Cf. also Popper [1961] p. 3. (I am grateful to Alan Chalmers for locating these two references for me.)

30. I try to do this in somewhat more detail in my [1983].
31. Marx-Engels [1975] 5:5 (Marx-Engels [1956] 3:7).
32. Marx [1976 – 81] volume 1, pp. 283ff (Marx-Engels [1956] 23:192ff).
33. Marx [1973] p. 101 (Marx [1953] p. 22).
34. Marx [1976 – 81] volume 1, pp. 293ff (Marx-Engels [1956] 23:200ff).
35. Cf. Godelier [1978].
36. Marx [1976 – 81] volume 1, pp. 166f (Marx-Engels [1956] 23:88).
37. Marx-Engels [1975] 5:54 (Marx-Engels [1956] 3:38).
38. Marx [1970] p. 20f (Marx-Engels [1956] 13:8f).
39. Marx [1976 – 81] volume 1, p. 91 (Marx-Engels [1956] 23:12); volume 3, part III. Roy Bhaskar's book [1978] argues that all law statements, including those in the natural sciences, are statements about tendencies.
40. Cf. Marx [1976 – 81] volume 1, p. 798; volume 3, pp. 339ff, 355 (Marx-Engels [1956] 23:674, 25:242ff, 256).
41. Marx [1976 – 81] volume 3, pp. 1019ff (Marx-Engels [1956] 25: 886ff).
42. *Ibid.*
43. See especially Marx [1976 – 81] volume 2, chapters 10, 15, 25 of the English edition, chapters 8, 13, 23 of the original.
44. Marx-Engels [1975] 6:482 (Marx-Engels [1956] 4:462).
45. Cf. Raymond's recent book [1982] which came into my hands only after I had finished this paper.
46. On this cf. Althusser [1971] p. 156.
47. But Claudin shows in his excellent book [1980], especially pp. 35ff, that *The Manifesto of the Communist Party*, written 1847 – 1848, has strong tendencies to inevitabilism.
48. Gramsci [1971], p. 336.Cf. also Korsch's interpretation of the theoretical features of the Second International in his [1970] pp. 57ff, 116f, and the remarks in Bettelheim [1978] p. 550 on the political significance of 'evolutionism' in Stalinism.
49. Gramsci [1977] pp. 34ff.
50. See e.g. Popper [1966] volume 2, pp. 369ff and especially pp. 381 – 383. For remarks on the 'field' of Popper's own thought see Krige [1978], Ravetz [1981] and D. Bloor [1978] chapter 4.

11. POPPER AND GERMAN SOCIAL PHILOSOPHY

Robert Ackermann

It will be assumed here that the reader of this essay is familiar with the general outlines of Popper's philosophy, but not with the controversies involving Popper's philosophy in contemporary German philosophy. These controversies have indistinct outlines for those who are restricted to the literature that is currently available in English. A reader of the Library of Living Philosophers' volume on Popper, for example, will find no mention of the most important controversies.[1] Against this background, the following essay is intended as a modest guide.

There are a number of philosophers whose original work was written in German, but whose sustained influence has been limited until recently primarily to English-speaking philosophers, a situation related to academic movement during the second world war. Any detailed attempt to deal with this phenomenon in general would involve, for example, both Wittgenstein and Popper[2]. Both of these last mentioned philosophers have been growing in importance in German philosophical discussions since the war, probably partly because of a readily understandable post-war influence of English and American analytic philosophy, an influence that has also produced an active literature of resistance. When Popper is the focus of discussion, an underlying or hidden agenda may involve the willingness of German philosophers to accept this influence, *particularly* in the area of social and political philosophy[3]. Popper has explicit professional defenders in German philosophy, notably Hans Albert, and Popper's philosophy is favorably invoked in the political defence of social democracy[4]. The possibly still growing importance of Popper's philosophy, based on this support, is also indirectly stimulated by negative philosophical treatment. To bring this varied literature into any focus, it is necessary to say something about the shape of German social philosophy in contrast to its American and English counterpart.

Currie, G. and Musgrave, A. (eds.), Popper and the Human Sciences.
 ISBN 90-247-2998-X

Popper's general methodological views are oriented towards, and compatible with, the sort of quantitative empirical research that is heavily represented in typical English and American departments of sociology and political theory. In Germany, this tradition has made a relatively minor and relatively recent contribution to social philosophy. Because of the conceptual division of German academic subject matter, however, history counts as a science with possibly special methodological problems, and so does sociology in a form that would be regarded elsewhere as general philosophical theorizing that is relatively independent of the constraints of small scale empirical data. Rather intense conflicts have developed between Popper's claims in these areas and German historians and sociologists with philosophical interests, conflicts that are not so strongly and extensively marked within the empirically and analytically oriented English speaking philosophical world[5].

Two major Popperian texts, *The Poverty of Historicism,* and *The Open Society and Its Enemies*, are the central arguing points of the conflicts that have been alluded to. Unfortunately for stylish organization, these texts are *not* independent, since Hegel and Marx are attacked by Popper in both books, and this has produced what might be expected: automatic counter defences by vested interests. A certain style of Marxist defence can be readily anticipated, and is readily available[6]. I shall try here to deal with more subtle Hegelian and Marxist rejoinders.

There *is* a historicist tradition among German writers of history that is very difficult to connect with positions outlined in Popper's book on the poverty of historicism. Nearly all the commentators on Popper's book have agreed that it is difficult to discern precisely what Popper means by historicism, since not even Marx seems to have been a historicist (or historist) as these terms are used by Popper[7]. In many cases, historicist historians were quite consciously opposed to Hegelian systematizing, feeling that Hegelian systems tended inevitably to lose too much of the historical detail required to understand historical epochs[8]. Of course this is not compatible with what Hegel seems to have *meant* by the possibility of concrete history, a history intended to involve a philosophical representation of an era that would lose no essential detail whatsoever, but fit a complex welter of material into an intelligible framework. The historicist historians complained that, Hegel's philosophical intentions notwithstanding, Hegelian history simply lost too much detail.

Popper would not, in terms of his own methodological reflections, share historicist criticisms of Hegel. For example, the historicists seem typically to have believed that there are nations or other natural human groupings which can only be understood as unique entities, and then only after a complex sympathetic immersion in vast quantities of material, leading to the result that these groupings cannot be comparatively evaluated by an external standard. Because of this, the historicists may seem to be relativists to some, although they did *not* hold that the values instantiated in different cultures were equally acceptable precisely because there would be no standpoint from which this judgment could be made[9]. Clearly Popper's position on rationality, historical explanation via general laws, and the myth of the framework would be incompatible with historicism in this sense, and Popper's insistence in *The Open Society* that increasing openess is generally desirable would also be incompatible with historicism, but these matters are not brought to bear on the specific works of major historicist writers in this tradition, such as Leopold von Ranke. Popper has simply not engaged the German historicist school of history at a level that would make for meaningful debate.

If we turn to Popper's presumed major targets, Hegel and Marx, a similar situation exists. Neither Hegel nor Marx is recognizable by sophisticated scholars in the portrait of the historicist that Popper offers. For example, many investigators have reached the conclusion that neither Hegel nor Marx thought themselves to be forecasting the future with the precision of astronomers, and that both were concerned rather to understand how the present had come about, along with hazarding some guesses about possible paths for future development[10]. Further, although Popper indicates explicitly that he is aware of divergent strains of Marxist thought, these are not contrasted separately to his methodological strictures, and the result of embedding Hegel and Marx into the historicist tradition tends to produce a conflation of their views and a simplicity of outline that is difficult to accept as a statement of the opposing positions. One can be Hegelian without being a Marxist, and Hegelians need not belong to any tradition that attempts to locate the essences of actual social groups as a clue to their futures[11]. Marxism may also be construed as a theory of ideal social types, actual societies containing contradictory elements whose resolution need not be determinate. These complexities blunt the thrust of Popper's criticisms.

Another important fact that Popper seems not to realize is that

Hegel is frequently the philosophical source of humanist versions of Marxism that are opposed to mechanical forms of Marxism of the sort that Popper seems especially to loathe[12]. This theoretical possibility of conflict within Marxism could not be elaborated within the one-dimensional caricatures provided in Popper's account. All of these matters seem sufficiently established by a series of careful studies of Popper's scholarship in this area[13].

The historicist position that Popper constructs in *The Poverty of Historicism* is apparently polemically motivated. If Popper were right that historicists had to be essentialists and forecasters in his sense, then *The Poverty of Historicism* would have settled the debate: it would not be rational to be a full-fledged Popperian historicist. Perhaps this is why there do not seem to have been many Popperian historicists. Popper holds views that are incompatible with the views of nearly everyone who has been identified as a historicist, but refutations of these positions would have to descend to historical details that do not engage his attention in his text, details that would make the arguments subtler and less conclusive. Popper may be hoist here with his own petard, since it is his picture of historicism that is essentialistic, but there may be no essential core to the various historicisms. Problems with Popper's treatment of historicism will recur below.

Popper seems to feel that a commitment to scientific social science requires the denunciation of historicist thought because of the awful price which persons living in Marxist countries have paid for social programmes based on Marxist doctrine[14]. There are a number of problems with any attempt to link historicist thought and totalitarian barbarism. To begin with, the attempt to derive denunciation from methodology, that is, to argue that historicism entails a form of society that is incompatible with Popper's concept of the open society, involves a patent logical error. Popper's concept of openness tends to beg the question because of a conflation of openness with individualism. If an open society, in contrast to a closed society, is characterized by the availability of more alternatives in its planning for the future, there is no necessity that the choice among options be based on liberal voting patterns and piecemeal engineering technology rather than on some collective process that might choose some strong reform option after resolution of conflicting interests[15]. At a sufficiently abstract level of theory, all sorts of social systems might be open (modern) *and* humanist. In practice, the effects of corruption demonstrate that inhuman barbarity is associated with all forms of self-perceived social structure[16]. There is something wrong with blaming Marx and Hegel for what goes on in countries with Marxist regimes, unless one is willing simultaneously to hold Christ, Democritus, and Thomas Jefferson responsible for what happens in various 'democracies' around the world, or Popper himself responsible for certain boring lectures on methodology said to have been given at various English universities. Why it is that certain texts are alluded to as constituting a sacred legitimating corpus in various societies is an interesting problem, but is is clear at the outset that there is no causal connection in typical cases between such texts and the activities in the relevant societies. The alternatives are not reason and violence. Violence may be explained by reason, or at least by deduction from given values, and non-violent resistance may not be reasonable. Context is too important to permit sweeping judgments here. Pressing the open society and piecemeal reform is without content when we are confronted with such possibilities as nuclear warfare, and no connection between these concepts and what to do in such a particular context.

What has just been said is becoming more pertinent in a time when Popper himself is cited in a justificatory fashion in German political debates. Popper's philosophy, stressing stringency of criticism and elaborations of alternatives as necessary for the advance of science, turns easily into a conservative instrument in its political employment[17]. Where reforms are suggested, the Popperian stress on piecemeal social engineering can be used to urge caution. The effect is to stabilize the status quo and forego any significant risks that may be associated with change. Violent shifts from one theory to another, forced at times within science by instransigent facts, are damped down when Popper's theory is imported into the social arena. This is, of course, related to Popper's arguments for preserving the autonomy of science. Let us assume that the Popperian conception of methodology is acceptable for explaining scientific progress where scientists are perceived as a separate community dedicated to the mutual understanding of nature[18]. There are many values involved in supposing the scientific community to have the independence and common goals it must have if Popper's methodology is to be plausible, values which are more likely to have been instantiated in the relationship between the relatively individualistic and autonomous science of the nineteenth century and its social setting, than they are to be instantiated in the twentieth century. The same social structure, imagined sympathetically in the larger society, runs into the difficulty that no common value in the larger society may correspond to the mutual goal of understanding nature among scientists[19]. Rather, irresolvable conflict of interest may predominate, a situation not conducive to the successful operation of the polite normative rules of mutual criticism that can be urged on the scientific community as a means of fuelling progress. *Progress* points in too many directions for too many people in the wider society for debate and mutual criticism to resolve the essential issues. Covert action to achieve perceived interests may often seem a perfectly rational mode of operation. Further, the sheer costs of modern science have made the question of the independence of science moot. Pure epistemological steering has had to yield to request for public funds, a request whose resolution inevitably escapes epistemological resolution within the domain of pure science[20].

Popper's attempts to transfer methodological lessons to social policy may not survive careful philosophical criticism. Necessary links between historicism and barbarism, and between the open society of scientists and the wider society surrounding science, simply have not been, and perhaps can't be, forged. The effect of this is that Popper has proved most congenial in the polemical defence of social democracy, where many of the challenging features of his scientific methodology appear in a transmogrified, and essentially conservative, form. A friendly commentator might only observe that a great deal of work is required if the social prescriptive side of Popper's polemics is to find a permanent place in German social philosophy, work that would need to take the direction of filling in the chasm between the pure theory and detailed political realities. In such a case, Popper's theory might be extended to a social philosophy with argumentative power against the unpersuaded.

One reasonable guess as to the source of Popper's claims about historicists is the observation that no *single* historical process seems a sound basis for guessing scientifically at the future of the process. Such a guess would only sketch a trend, rather than a law, since it would apply only to the process from which it was derived, and would not be general enough to be tested by other processes[21]. Problems with the scientific status of evolutionary theory are connected with some such observation in the eyes of some methodologists. Popper believes that we can make predictions, or guesses, about the future, but in science these are based on general laws that can be severely tested, and hence earn high corroboration. If we return to Marxism and historicism for a moment, it seems clear that Marx was not a historicist in the relevant sense, but took himself to be developing a science of historical explanation that would apply to a wide range of human historical sequences. Thus he was aiming at the generality of science, and the problem becomes one of deciding whether Marx thought he could make infallible predictions. A great deal of debate about Marxism as a possible science can be short-circuited with the observations that there is no need to assume infallibility in Marxist discussions of the future[22]. Some current economic forecasting which is committed to continuing economic growth, ensuring economic and social stability, and finding an economic system which transcends awkward features of both capitalism and socialism can be historicist (in Popper's sense) while developing a piecemeal social technology designed to monitor and control projected growth[23]. Quite specifically then, historicism can seem to arise outside the Hegelian tradition, and it need not be linked to infallible forecasting. Popper's blanket condemnations do not seem to hit a very well defined target.

What we shall do now is to turn to the more permanent influence of Popper that may be expected to remain in German discussions of social science methodology as a result of the positivism dispute. Popper's social science methodology is intended to bear a close correspondence to his natural science methodology, so that Popper is not sympathetic to methodological conceptions which see a great difference between the two clusters of specific sciences[24]. In particular, Popper avoids the division based on the conception that infallible prediction is possible in natural science but impossible in social science because he holds that all of our knowledge is conjectural, and problems with theoretical indeterminism can provide no dividing line between infallible and fallible conjectures because they already infect the core discipline of classical physics[25]. Explanation proceeds everywhere by deduction from (conjectural) laws and boundary conditions, and science is everywhere motivated by the desire to solve specific problems that are posed by everyday (and scientific) observation[26]. The major requirement of methodological individualism in the social sciences directly avoids pitfalls of historicism because all societies are broken down into individual human beings who are assumed to be rational and to respond (typically) in similar ways to set external conditions no matter what their cultural setting. Psychological states fall out as irrelevant. Humans do what they do because of the logic of the situation, and the logic of the situation forces their behaviour along certain lines among the alternatives available in the situation because of their rationality[27].

The social sciences study these situations and the consequences of the actions available in them. It is especially pertinent for the social sciences to study the consequences of certain actions which might not be noticed by the relevant actors, and would constitute unintended consequences of action were the actions to be performed in this state of ignorance. Explanation only requires laws which apply to all situations (and all human beings), so that their testing is possible because they can be tested for each individual, and tested in various instantiations of elementary situations. History then becomes the social scientific study of the past. On Popper's view, the laws about human beings used in explanation will be subjectively trivial, or nearly trivial, since their instantiations would be so numerous as to constitute part of our (possibly explicitly unacknowledged until formulated) understanding of human nature[28]. Popper clearly lays down a conception of methodology which is incompatible with the features of historicism that he wishes to combat.

In the writing of history, as well as in the practice of social science, Popper stresses the necessity of a point of view to narrow the choking complexity of actual events into a manageable scenario. The focus of a problem in contemporary investigation may be replaced in historical narrative by a point of view. An implied contrast is to the tendency of historicist writers to tell history as it actually was[29]. What Popper's account misses is the fact that although historians may work from a point of view, they often treat their interpretations as in conflict[30]. The conflict between existing interpretations, or their inadequacy to newly discovered fact, is often the starting point for new historical writing. Perspective need not be chosen out of individual values; it may be chosen to illuminate what we want to regard as the truth about some historical moment[31]. Surely Popperians would want to regard the history of science as written from the viewpoint of conjecture and refutation as more revealing of actual mechanisms of scientific progress than alternative accounts would be: otherwise the claims made for the Popperian conception of methodology would have no interesting connection to scientific practice. Historians feel similarly about their craft. Where they shy from Hegelian systematizing they will still feel that modest comparisons between epochs, or across cultures, may be revealing of mechanisms of social movement, even though (general) laws of history cannot be formulated. This thrust towards explanation assumes that values can be thrashed out, and at least some point of view established as superior, more revealing of the causes of social change, than others. Popper's conception of methodological unity requires him to hold that this level of historical explanation cannot be regarded as objective knowledge except insofar as it can suggest generalizations which are falsifiable. One rejoinder is simply that historical knowledge *is* knowledge in the form of detailed factual synthesis, and that disagreement about the significance of such syntheses is subject to eventual resolution among historians, whereas the conception of methodological unity would require history to be written at a level of general conjectures and refutations not very often attainable in practice. The continuing impact of Popper on historiography is likely to be minimal without some accommodation to the practice of writing history.

We can now turn to the more specific positivism dispute, a dispute likely to have a lasting impact on discussions of social science methodology. This dispute in its modern form can usefully be regarded as triggered by a conference at Tübingen in 1961 on the logic of the social sciences at which Popper and Adorno presented related papers. Following the conference, a series of papers by Popper and Albert, on the one side, and Adorno and Habermas, on the other side, have continued the dispute[32]. It seems best, however, to regard this dispute as the contemporary form of a deeply rooted argument in German sociology which goes back to the nineteenth century, before a time when economics and sociology had assumed their current academic form[33]. The debate originally involved contrasting reactions to the *failure* of the optimistic predictions inherent in Smith's and Ricardo's early theories explaining social progress in economic terms. One approach was to look for better theories along the same lines, that is, better and more powerful theories incorporating more empirical variables linked in a more precise fashion. This path leads ultimately to modern neoclassical (and other varieties of formal) economic theory. Against this was ranged a variety of thinkers opposed to the abstraction involved in making economism the sole explanatory spring of society. These thinkers sought no grand theories, but merely hoped to come to grasp local or national economies in their full complexity, and to make useful suggestions for their improvement based on this detailed understanding[34]. These latter economists are in some sense heirs of the historicist tradition in historiography, something that should be nearly obvious in the light of previous discussion. The so-called value dispute arose when Max Weber proposed, shortly after the turn of the century, that the *scientific* study of society should be value free, leaving suggestions for social policy based on large value questions to a social politics which could not assume a proper scientific form. An extremely tangled history has followed this suggestion, tangled particularly since Max Weber's views are widely regarded as allied with positivism in Germany, but as implicitly historicist in England and the United States. There are, as a result, two Max Webers to be considered[35]. Weber's insistence on rationality as the touchstone to explaining human behaviour is positivistic, but he also held that a process of intuitive understanding is required to grasp agent's situations and the alternatives open to them. This fact is undoubtedly basic in accounting for the ambiguous nature of his identity in the two traditions. When Weber's methods are restricted to relatively small scale subject matter (much smaller than the subject matter attacked by Weber in his own work), so that intuitive understanding becomes relatively trivial, Weber seems to stand in an important place which is re-occupied by Popper in its essentials in the positivism dispute. A most important aspect of this is Popper's insistence on the clean separation of fact and value, and the restriction of scientific investigation to value free hypotheses[36]. What this means in practice is that the social scientist should trace out the factual consequences of certain situations, consequences that would be the same no matter whether the investigator were to be a social democrat or a communist. These scientific linkages would then be data available for social decisions.

Critical theory, as represented by Adorno and Habermas, is precisely opposed to the idea that fact and value can be separated in a way that allows a value free science to gain any useful insight into social structure. For philosophers in this tradition, past and present societies have presented themselves to their members in an illusory form, recently as a set of individuals bearing only external relationships to one another. In this form, the reality of society seems to be comprised of individual persons, tokens of money, objects manufactured by human beings for the use of other human beings, and so forth. The locations and movements of these building blocks of society can be plotted in the space and time of physical science. This is the appearance which is studied by an abstractive economics and sociology. Actual society is hidden behind this appearance for Adorno and Habermas, and society must be grasped through its appearance. There are differing accounts of how the reality of society can be studied in critical theory, but on all of them, when the reality of society is grasped, the *appearance* of society will no longer be the same. It will now be seen *as* appearance, and its intrinsic historicity (the fact that it represents something that must change as a result of intrinsic contradiction) will become obvious. In Marx, this route is marked in *Capital* by the early examination of the commodity in the capitalist market, something that may seem a simple physical object with well defined properties, but something that turns out to be a mysterious appearance of somewhat uncertain and mysterious properties when critically examined. Critical theorists have seen the clue to underlying social structure in some of the art objects produced in societies, or in patterns of communication within societies, but these complexities cannot even be sketched here.

Popper and critical theorists meet, to be sure, on somewhat different planes. For Popper, economics and sociology start with problems. Why is such-and-such a train running late? Why is the price of tomatoes what it is? And the answers will involve the activities of individuals within the context of known social institutions. Perhaps the engineer slows down to wave at his family when he shouldn't, or various farmers just got tired of planting tomatoes because of an unusually hot spring, and so fewer plants were set out than usual. Now these are explanations from one point of view, and they may help to solve the particular problems of getting the train back on schedule or changing the price of tomatoes. But this can happen only within a social structure that is assumed to remain fixed in its larger outlines. Critical theorists are not interested in these specific questions within fixed social settings, but rather in questions about the nature of the *link* between transportation *and* vegetable pricing in total social structures. They may be interested to know whether this social structure is fascist, and how that is related to such links, and to the possible future of the society. As they move to such general questions, their explanations become historicist in the sense that they deal with unique situations, attempting an understanding of these situations in their full complexity and concreteness without the availability of suitable general laws of society. This is just to deny that there is a form of human rationality that is constant across social structures at a suitably concrete level of investigation.

Now let us turn to the Popper-Adorno confrontation. Popper had prepared for it explicitly by presenting twenty seven theses concerning social science knowledge, asking that Adorno accept or reject them. This Adorno refused to do. The usual interpretation is that Adorno and Popper simply talked past one another[37]. A closer look suggest something different; that Adorno would not discuss the theses because their language created a space for discussion within which his own views could appear only in a distorted form. This can serve as a presentiment of the structure of their disagreement. For Adorno, it was as though a doctor were asking questions about the symptoms of a disease whose cause and nature could only be understood in a much wider framework embracing the historical and social position of the diseased body. Popper's initial theses suggest that we have a great deal of knowledge (although we are also ignorant), and that social scientific knowledge should be attacked by formulating problems and hypotheses, and then looking for falsification, exactly in line with scientific methodology elsewhere, except that the rationality postulate may be involved. This view suggests that we are currently in a position to determine what problems are significant and worth attacking, and that we can see what evidence would falsify social science hypotheses. This raises a direct confrontation with critical theory. Critical theorists see social problems as mediated by society, so that they cannot be separated and separately attacked in an arbitrary temporal order (the order in which they occur to us). No matter how independent two phenomena may seem to be on first appearance, there can always be some strong connection between them which is established by a social linkage that is perhaps not conscious to any particular actor. A synthesis of all problems conceived of in terms of our current outlook would no doubt be contradictory, if only due to the conflicting interests of various groups and/or classes, and some problems would turn out to be discardable or automatically solved in some rational order of solution. Piecemeal social engineering might produce an early warping of social direction that would later be irrecoverable, if only because of the limited scope within which the original questions was attacked. Popper's attitude is direct: roll up your sleeves and attack problems. Critical theorists urge reflection on wider contexts before hypotheses are formulated and data taken, so as to maximize the chances of taking permanently significant action. Adorno says this as follows:

> But the cognitive ideal of the consistent, preferably simple, mathematically elegant explanation falls down where reality itself, society, is neither consistent, nor simple, nor neutrally left to the discretion of the categorical formulation (Adorno [1976] p. 106).

These positions are not logically contradictory. One could be forced to consider wider and wider issues in pursuing piecemeal social engineering, until the conceptual framework and perhaps even societal relationships presupposed in the original formulation were called into question. And one might find that a consideration of the mediation of problems left everything intact except for the question at hand, for which the anticipated falsifying data were relevant. But the focus of piecemeal social engineering is conservative; it expects to solve problems satisfactorily within the current scientific and social framework. The focus of critical theory is revolutionary; it expects that society will require fundamental change before a coherent social science will be possible. Underlying the logical compatibility of the positions are views of contemporary society which force them into confrontation. Popper, in accepting the given structure and attempting to preserve it while pursuing modification, produces controversy within a framework where its resolution (by falsifying hypotheses) can be a joint task. Critical theory, on the other hand, doesn't offer the outlines of the future it envisions with sufficient clarity to permit unconstrained allegiance, and it provides no mechanism for resolution of disputes between critical theorists about the order of social science inquiry.

This substantial debate and current form of the old struggle over historicism is concealed in the literature, unfortunately, behind minor issues. Adorno identified Popper as a positivist, a charge that Popper met with umbrage. In Popper's terms, a positivist believes in a justificationist, foundational epistemology, while he, Popper, does not. Therefore he is not a positivist. Q. E. D. For Adorno, a positivist is anyone who accepts the appearance of society as social reality[38]. Popper's brisk way with identifying and attacking social problems convicts him of positivism in this sense of the term. The social role of positivism, from this latter point of view, is to assist only trivial changes in the current social structure, and in the long run this serves the status quo. Popper does seem to be a positivist in this sense, but the situation is tricky. Adorno's definition of positivism is not coherently formulable in Popper's terms. So much follows from Popper's strictures concerning the myth of the framework. But root differences in social intuition surface again. In Popper's view, values can be resolved through discussion and compromise, but where this fails, are to be resolved through the democratic process. For Popper, the alternative is a polity where values are resolved by force. Adorno sees this view of democracy as naive, and as concealing power relationships coded into such matters as media selection and coverage of election issues. Popper wishes to export scientific social structure to the surrounding society, in which case the problems could be solved as they are within science. But perhaps there are non-scientific social problems requiring resolution along different lines. Adorno envisages a polity in which the separation of fact and value underlying democratic voting is overcome, and values can be rationally resolved *in every case*[39]. From his point of view, it is more essential to work toward such a society than to preserve the power structures inherent in current democracies. The crucial role these differing conjectures about society play in framing the debate has not been much discussed, and here it is noted without an attempt at adjudication. Between these conjectures there may be (ultimately) no compromise, and Adorno and Popper certainly did not find one.

If the positivism dispute represents irreconcilable conflict with Adorno's critical theory, Popper has more recently been treated with new interest in the eclectic syntheses of Habermas and Apel, and there are signs of a rapprochement with these versions of critical theory as they turn to different conflicts than those which had occupied Adorno[40]. Apel and Habermas have produced emancipatory social theories based on notions of complete and undistorted communication between members of society as a normative ideal. In their survey of previous social theories, Popper's plays an important role because it is the only well known analytical philosophy of science to have placed an emphasis on free and open mutual criticism (and hence communication) in the scientific community, as well as the fact that it is the only one to suggest that this might be a model for handling progress in larger societies. From this perspective, Popper is elevated from association with other positivists by his social impulse and sensitivities[41]. Other positivists attempted to ground science in the logic of theory and evidence, but Popper recognized that progress could be accelerated by the widened insights of mutual criticism. Here Popper is allied with Mill, both supposing that open and free discussion is the only route to truth when a foundational epistemology is not available with sufficient scope to provide the individual with a basis for locating truth. Open criticism comes close to distortion-free communication when combined with the normative goal of honesty.

If Popper is extended to allow the possibility of objective resolution of value conflict, he can almost be developed into consilience with hermeneutical approaches. In fact, of course, Popper remains a classical liberal of a well known variety, and a democratic theorist who fears coercion in value resolution. But Popper's liberalism leaves no room in his scheme for value ideals other than freedom, justness, and equality, values which are consonant with his liberalism[42]. They are subjective norms in that each person has the right to his or her value opinions, and no notion of the responsibility of the person to defend value choices plays a role in Popper's system. No norms independent of individual human desires can be fitted into Popper's system, *including* norms of rationality. As is well known, Popper describes the choice to be rational as itself an irrational choice, and restricts himself to developing the consequences of this irrationality. Clearly, Popper's system can't close itself without a historical dimension, and without some method of rationally gounding values which transcend individual desires. The attempt to rationally ground values has been a central core of the idealist response to Kantianism in the history of German philosophy. We have here the same structural problem that many philosophers have found puzzling in the relationship of Kant's theoretical philosophy to his practical philosophy. The rational closure of Popper's Kantianism seems to call for a Hegelian turn, against Popper's will, and this seems to be the major constructive result of the recent confrontation of Popper's philosophy with the German tradition of idealism. This is not to say that Apel's or Habermas' version of the freely communicating society with its normative overturns of value resolution is the natural terminus of Popperian reflection, but merely that movement in this direction would follow a familiar philosophical dialectic in attempting to find a philosophical system with no self-admitted irrational planks.

This direction might have been more apparent than it has been if methodological individualism had been more closely related to economic theory. Methodological individualism is an attractive idea for philosophers who conceive of society as an assemblage of freely acting individuals, as Popper does. It can therefore work without strain, even where it is only a methodological strategy, within various historical and sociological contexts. But Popper has himself stressed the importance of institutions and their relative independence from human beings in the short run. Institutions may well transmit individual human actions into vast unintended consequences elsewhere. What seems not to be noticed in the Popper camp is that neo-classical economic theory (as well as Marxist economic theory) sees economic relationships as penetrating the entire fabric of society. A price change, or a supply change, even in some relatively small market sector, will instantaneously affect the entire society, that is all other prices, or an entire schedule of demands. Economic theory then drives us to the conclusion that the social effects of individual action may well be holistic, and this is independently of Marxism, and not obviously compatible with the presupposition of piecemeal social engineering that small changes will have small effects. The theory explicitly confronts the possibility that catastrophic effects can follow from extremely modest causes in the right circumstances. Methodological individualism suggests a loosely jointed society of independent individuals, but when it is coupled with a theory of institutions, it can lead quite quickly to a holism of its own. In the end, then, the choice is not a simple one between measured change and uncontrollable disruption, or between piecemeal social engineering and revolution, it may be choice between holistic changes, even when power relationships, class struggle, and misinformation are not considered. Rather than defending an abstract methodological scheme, the status of economic theory in all of its variants calls for Popperian attention to the details of social problems, and to a developed notion of rationality (if not Habermas' or Apel's) that can deal more effectively with value conflict in the world as we find it.

Notes

1. Schilpp (*ed.*) [1974] contains thirty three articles on Popper's philosophy, with replies by Popper. Very few of these are of concern to our topic, and such important issues as the positivism debate are not mentioned anywhere in this volume. Ackermann [1976], Johansson [1975] and O'Hear [1980], the available general studies of Popper in English, discuss relevant topics only briefly. Jarvie [1972], which contains a good discussion and exposition of Popper's conception of social science, is completely silent on the positivism controversy. Marcuse [1972], which is directly relevant, is not mentioned very often in the English literature.
2. Popper and Wittgenstein were both Austrian, and not German, in origin. Could this matter in any way? See König [1959], for illumination.
3. Heller [1978], provides an insightful look at the political significance of postwar philosophical disputes.
4. For a collection of articles sympathetic to Popper's social theories in the German context, see Lührs et al. (*eds.*) [1975]. (There is also a second volume of discussion). For Albert, see articles by him in Adorno [1976], Giddens [1974], Lührs *et al.* (*eds.*) [1975], and Topitsch (*ed.*) [1976], plus their bibliographies. An interesting analysis of neo-classical economic theory from a Popperian point of view (in English) can be found in Albert [1964].
5. There is no important, academically central, and widely respected Hegelian or Marxist tradition in America or England, so that Popper's social philosophy has been discussed there in an atmosphere where the outlines of Popperian social philosophy are nearly truistic, and where technicalities are the normal focus of objection. Popper's attraction is much more easily felt here than it has been in the more varied European setting.
6. See Cornforth [1968] for an example.
7. For discussion, see Freeman [1977] and Bayertz et al. [1977]. The latter is much more thorough in terms of sources. (In the first part, Bayertz discusses Popper's epistemology. In the second part, Schleifstein discusses Popper's social philosophy.)
8. See the definitive discussion in Iggers [1968].

9. *Ibid.* presents a sound and detached survey of historicism. Historicism is opposed to the natural law tradition which stresses a universal human rationality. Because of this opposition, historicists hold that system and general theory is out of place in economics, history, and sociology. The common properties of nature over time that underly the progress of the natural sciences are not thought by historicists to be met in social nature, where achieved understanding is always threatened by shifting social structures. Donagan's discussion in his [1974], Popper's reply later in the same volume, and Lee et al. [1954] are also of interest in this connection.
10. See Albrecht [1973], Freeman [1975], and Bayertz et al. [1977].
11. Albrecht's [1973] is a careful consideration of these issues in terms of non-Marxist Hegelianism that easily brings to light Popper's superficial Hegel scholarship.
12. For a brief introduction to this issue, see Jacoby [1981], especially pp. 23 – 24.
13. See Albrecht [1973], Bayertz et al. [1977], and Spinner [1978] Volume I. (A second volume is to have appeared. Volume I contains an excellent bibliography.)
14. See Popper's dedication to his [1964].
15. An extensive historical discussion of Popper's notion of an 'open' society is to be found in Spinner [1978], a discussion leading to this conclusion.
16. Marcuse notes this in his [1972]. Some philosophy departments can provide small scale, but ready to hand examples.
17. Marcuse in his [1972], p. 200, points out that the political context of Popper's philosophy grows out of its methodological content. The revelation of the political in the methodological cuts against the postulation of a neat fact/value dichotomy. Spinner [1978], contains an extensive discussion of Popper's use in current political discussions in German, with specific references. See especially pp. 39 – 56.
18. This point has been made repeatedly by Habermas. His [1973] contains the most often quoted discussion.
19. This point has also been argued extensively by Habermas, and also by Spinner. See for example, Spinner [1978], p. 554.
20. Ferber [1976] argues forcefully that the only relevant question is which values science will help to realize, not whether values are involved in scientific work in a substantive way.
21. This is very succinctly and convincingly developed in Miles [1975], p. 4.
22. Albrecht [1973], pp. 171 – 175. Marcuse [1972] points out that not all of the institutions of a society need be analysed in a holistic theory, only essential or most prominent institutions, and these latter are all that a totalitarian need control in order to achieve dominating influence in a society. In current Poland, for example, the mere cutting of public forms of communication allows strong totalitarian control (at least in the short run) without a full historicist analysis of the society.
23. Marcuse [1972], p. 199, points out that cyclical theories can be developed which may posit historicist inevitability grounded in an empiricist epistemology. Miles [1975] suggests that W.W. Rostow and others are best described as empiricist historicist social engineers in Popper's framework. From this point of view, Popper's target was much too narrowly focussed in his [1964].
24. A prominent line of argument against Popper's general methodological scheme is due to Wellmer in his [1967], pp. 44 – 61, who argues that Popper's psychology, in making expectation basic to avoid Hume's inductivism, threatens the existence of the notion of independent reason that Popper needs for the development of critical rationalism. Inborn expectations cannot be the same as rational conjectures, but the presence of rational conjectures is not subject to falsification because of the weak link between them and behaviour. This objection amounts to saying that Popper's psychology is an *a priori* conjecture, and not a scientific theory. Wellmer suggests that Popper may need to become more Kantian and add a notion of the transcendental *a priori* to his philosophical basis in order to avoid *ad hoc* foundations for the development of critical rationalism. Wellmer's critique would require expansion in view of Popper's later emphasis on language as permitting the objective statement and critique of hypotheses. Critique, for Popper, is primarily a matter of sharpening hypotheses so that empirical refutation becomes possible under sharp conditions, and then seeking that refutation. World 3 implies a sort of self-transcendence in critique, in that the *community* of scientists can participate in critique, but not a self-transcendence that escapes the logic of falsification. A self-transcendence that can

escape the constraints seemingly involved in dealing with a proposed hypothesis, which critical theorists find implicit in language usage, and which they see as permitting a critique of hypotheses as formulated within a (scientific) false consciousness that tacitly accepts the validity of current scientific practice and its relationship to surrounding society, is not compatible with Popper's stricture against psychologism. Whether such a self-transcendence exists, of course, is at issue. This is relevant to the discussion of value resolution which occurs below. Popper's general views on the philosophy of science are discussed in the relevant German literature, see for example Bubner [1981] and Seiffert [1975]. (This appears in two volumes with an additional volume surveying relationships between Marxism and bourgeois philosophy of science.)

25. Ackermann [1976], pp. 129–140, contains an introductory account.
26. Popper seems not to have discussed very carefully the methodology that may be needed for investigating the structure of his so-called World 3, as opposed to the motivation for introducing it, which is to avoid subjectivism in the philosophy of science. See Ackermann [1976], pp. 54–57, for a discussion and references. As there is only one World 3, a hermeneutical or dialectical method may be required for its development, a fact noted, for example, by Schupp [1975], p. 270.
27. Jarvie [1972] contains a useful elaboration.
28. Wilkins [1978], pp. 32–40, discusses this aspect of Popper's work, and notes that Popper provides two examples of historical explanation in *The Open Society*, the division of Poland in 1772, and Caesar's decision to cross the Rubicon. The distance of Popper's treatment of these examples from good historical practice will be obvious to the reader.
29. *Ibid.* p. 103, makes this point.
30. See Schupp [1975], pp. 167–185.
31. *Ibid.* especially pp. 262–269. Skagestad [1974] contains an interesting critical discussion of Popperian remarks on history from a completely different perspective.
32. There occurred a spin-off debate in German pedagogical theory. In this debate, it was argued on the one hand that pedagogical assertions should be falsifiable, and on the other that they could not be falsified on the scientific model because they contain in all interesting cases important value judgments. Art history, for example, cannot receive a curriculum and be developed without a prior judgment as to what significant art is. This debate is close to the debate about history discussed above. For a summary and bibliography, see Beck [1974].
33. Lorenzen [1970] argues that the dialectic (especially in Hegel) is an attempt to defeat or at least sublate Enlightenment values and save the moral and philosophical content of Christianity. Reason is given to us by God as our guide, but the Enlightenment displaces the study of reason to a study of nature as the central philosophical task. The heirs of the Enlightenment see the attempts to locate objective norms as sheer romanticism; for them the world no longer has a necessarily rational structure. Lorenzen attempts to ground the objective values needed for science compatibly with the Enlightenment tradition. For an interesting and incisive discussion of the history of the struggle between dialectical and Enlightenment values, see Thiel [1972].
34. Miles [1975], p. 14, argues that Marx was attempting to found a science that would escape the parochialism of historicism. This is also the staple of one French school of Marx interpretation, but the whole project depends crucially on what *science* is taken to mean, and the issues are too complex to be developed in this paper. Wellmer [1974] can be recommended as a starting point.
35. König [1959] suggests that social life in Germany (due to its Lutheran traditions) has always been viewed as having lesser status than some form of (inner) personal development. Social life is inherently self-alienating from this perspective. This fact may play some role in the background attitudes encountered in Hegel, Marx, and representatives of critical theory. By contrast, Weber was actively interested in social life and politics, a possible source of his ambiguous status. On Weber as a forerunner of Popper, see the introductory essay by the editors in Lührs *et al.* (*eds*) [1975], pp. 44–53.
    Huff [1981] presents a very interesting analysis of the methodology of the social sciences against the background of Weber's methodological opinions.
36. There are complicated issues here, since *some* values are obviously integral to science, particularly norms of conduct and standards of practice in specific fields. These norms have

to be filtered out by analysis before a fully accurate statements of the fact/value dichotomy can be supported. Albert has made major contributions to clarification in this area. See the sources mentioned in footnote 4.

37. See Popper's remark on p. 289 of Adorno et al. [1972], as well as the remarks on the discussion by Dahrendorf in the same volume. Giddens, in his introduction to his [1974], also suggests that no engagement occurred.
38. This suggestion is also made in Albrecht [1973], but not from the point of view of critical theory.
39. Following Lorenzen [1970], we can distinguish between right-wing romantics who are nostalgic for the past, and left-wing romantics who look for a better world in the future. Left-wing romantics tend to downgrade the importance of practical reason, and consider only pure reason as legitimate subject matter. This suggestion is complicated when set against the subtle positions of critical theorists about the past and the future. Adorno (and Horkheimer) in their [1972] seem to have held that the conclusions of practical reason are historically bound, and can only be drawn in certain social settings. Later developments in Habermas and Apel seem to promise universally valid normative consequences for theoretical reason.
40. See Apel [1980] and Habermas [1973], especially pp. 276 – 282 of the latter. The *difference* between Apel's and Habermas' development of ideal communicative societies as normative ideals for the discussion of truth cannot be developed here. For an introduction, see Bubner [1981]. Baier [1966] is an interesting discussion of the place of history in Habermas and in critical rationalism.
41. Throughout this paper, I have presented criticisms of Popper without consideration of possible polemical rejoinders. This effort has been directed at an exposition of the more serious critical positions that one cannot easily locate in the English literature. I have made clear elsewhere my enormous respect for Popper's achievements. (See Ackermann [1976], pp. vii – x.) The continuation of the positivism dispute seems an important item on the methodological agenda, and I believe that Popper's reappearance in the work of Apel and Habermas is an important testimony to Popper's permanent place in the philosophy of science, no matter how one may evaluate these systems.

    Popper's work has recently been widely cited in discussions of the possibility of a scientific literary theory. This impact lies outside our concerns here, although in the German context it can be regarded as part of the methodological discussion about social science. See, for example, various passages in Nassen (*ed*) [1979].
42. See Albrecht [1973], p. 148; Wellmer [1974], pp. 20 – 21; and Witschel [1971], p. 46.

12. SOCRATES AND DEMOCRACY

Richard Kraut

Was Socrates 'the champion of the open society, and a friend of democracy,' as Popper thinks?[1] Or was he an enemy of democracy, as many others have concluded?[2] I want to give Popper's thesis a limited defence: I believe that the Socrates we find in Plato's early dialogues[3] cannot be accurately described as an opponent of democracy. On the other hand, I wish to reject the arguments Popper gives for this conclusion. Furthermore, I will try to show that he overrates Socrates' credentials as a good democrat. If we want to call Socrates a democrat at all, he certainly was not the sort that Popper thinks he was.

For our purposes, it is important to observe the distinction between a *critic* and an *opponent*. An opponent of a form of government tries to change it, whereas a critic merely expresses judgments about its defects. In my opinion, Socrates was a staunch critic of Athenian democracy, but he did not oppose it. And although he severely criticized the government of his native city, he also appreciated one of its great merits.

1. Let us begin by looking at five passages in which Socrates attacks 'the many':

(a) In the *Apology*, he reminds the jury that he has not participated in the everyday political life of Athens[4]. His divine sign has warned against it, and rightly so: for he would have died long ago, had he taken part in politics. And then Socrates makes a general point about how corrupt the majority – any majority – is: 'Do not be angry at me for telling the truth: no man will be spared if he genuinely opposes you or any other great number (*plēthos*) and prevents many unjust and illegal things from happening in the city[5]'.

(b) Earlier in the *Apology*, Socrates cross-examines Meletus, and it

Currie, G. and Musgrave, A. (eds.), Popper and the Human Sciences.
 ISBN 90-247-2998-X

emerges that, according to this prosecutor, a great many Athenian citizens know how to improve the young[6]; in fact only Socrates corrupts them. To this, Socrates replies: 'Does that hold true of horses too, in your opinion? Do all men make them better, with the exception of one, who ruins them? Or it is just the reverse: only one can make them better, or very few, namely the horse-trainer, whereas the many, if they own and use horses, ruin them? Isn't that the way it is, Meletus, both for horses *and all other animals*?' (25a13 – b6, my emphasis). Most human beings are unable to give the young the moral education they need.

(c) In the *Laches*, Lysimachus asks two generals – Laches and Nicias – whether his sons should be trained to fight with armour. Nicias says yes, Laches no, and so Socrates is asked to cast the deciding vote[7]. But he protests: one should not decide such questions by numbers, but should consult an expert instead. 'It is by knowledge that I think one must make decisions, not by the greater number (*plēthos*), if one intends to decide well'[8].

(d) In the *Crito*, Socrates tells his friend not to care so much about what the many will think of them, if they make no attempt to escape[9]. We should value the opinions only of those who are wise, i.e. those who have developed a high level of expertise in a particular subject. For example, those who want to develop their bodies should seek the advice of a doctor or a trainer, and should disregard the views of the many. For if we slavishly follow the advice of the many, our bodies will be harmed. And similarly for moral questions: 'with actions just and unjust, shameful and fine, good and bad ... (we should not) follow the opinion of the many and fear it, (but) that of the one – if there is one who has knowledge of these things ...'[10].

(e) Socrates then goes on, in this same dialogue, to lay down the principle that injustice must never be done, even in return for injustice[11]. And he warns Crito not to accept this controversial point unless he really believes it: 'Crito, make sure that you don't agree to what is contrary to your own belief. For I know that there are few who do believe or will believe these things. Among those who are convinced of these points and those who are not, there is no common counsel (*boulē*). Rather, they must think the worse of each other when they consider each other's decisions' (49c11 – d5).

2. In the passages cited above, Socrates says that the many have three flaws: a large number of their moral beliefs are false; on many occa-

sions they act immorally; and they lack the ability to give their children a proper moral education. Does he think that these flaws can be, or someday might be, eliminated? Clearly not. For in (e) he says that few will *ever* agree with him that injustice must never be done; in fact, he is so convinced of this that he claims to know it. Now, if the many will always believe that injustice should be returned for injustice, then they will always act unjustly, if they are provoked and have a chance to respond in kind. In any case, (a) is most naturally read as a statement about the permanent corruption of the many: whoever opposes a great number and tries to prevent unjust and illegal behaviour shall not survive for long. Socrates is not merely saying that whoever opposed the democracy in the past didn't last long. Rather, he is making a sweeping statement about what inevitably happens whenever the many are opposed. Similarly, Socrates is saying that the many *will never* be able to give their children a proper moral education. If parents believe that injustice is sometimes justified, and if they will never change their minds about this, then they will naturally teach this to their children, who will in turn teach their children the same thing, and so on. Perhaps now and then an individual will overcome the effects of his early training; after all, Socrates has converted Crito to the view that injustice must never be done. But nonetheless, Socrates has become convinced that most people will learn false moral beliefs from their parents, will retain those errors throughout their lives, and will teach them to their children. That is what he means when he says in (b) that the many corrupt the young.

Now, someone could accept Socrates' low estimate of the many and still remain a proponent of democracy. One could argue that all adult human beings, regardless of their moral faults, have a right to participate equally in the making of political decisions. Or one might claim that participating fully in the political process of a country is a necessary ingredient of the good life, and that no one should be denied such a good merely because he has false moral beliefs. Alternatively, one might claim that false moral beliefs and immoral conduct are a universal and inevitable part of the human situation: since everyone has these flaws, they cannot be used to disqualify anyone from sharing political power. But even though it is possible – in these and other ways – to denigrate the many and still remain an advocate of democracy, there is no evidence that Socrates availed himself of this possibility. Nothing he says even remotely suggests that anyone has a *right* to vote or rule. Nor does he ever say that ruling is a good that every adult deserves to experience. On the contrary, he holds in Book I of the *Republic* that wielding political power is in itself unattractive: the only consideration that could lead a good man to rule is the fear of being ruled by those who are worse[12]. Furthermore, Socrates nowhere adopts the extreme position that the faults of the many are so widespread as to be universal. He and a few others have managed to free themselves from the common misconception that injustice is sometimes appropriate. At least to this extent, they are better than the many.

I think it is fair to conclude, then, that Socrates is a critic of democracy – rule by the many – for the simple reason that he has a low opinion of the many. Their false moral beliefs and willingness to engage in injustice will inevitably lead to bad laws, bad decisions, and official misconduct. There is never any suggestion in the early dialogues that the democratic process – the vote or the lot – is a fair procedure that we ought to value apart from its results. Instead, Socrates is evidently assuming that a way of making decisions is only as good as the quality of the decisions that usually result from that procedure. Decisions made by majority rule have all of the defects of the majority. And Socrates believes that these defects cannot be remedied[13].

3. Once it is conceded that, according to Socrates, democracy is inevitably a bad form of government, an important question still remains: did he think it possible to do significantly better? In my opinion, he was pessimistic on this score. He does conceive of the possibility that a city might be ruled by a few moral experts, and he is sure that such a city would be better governed by far than any democracy. But this is a merely theoretical possibility for him. For he has serious doubts about whether it is humanly possible to acquire the level of moral knowledge that would be needed to rule a city well. Socrates does not consider himself to be a moral expert, nor is he aware of any others who are. Furthermore, he doubts that human beings have the capacity to acquire the kind of knowledge that would be needed to rule well. And he thinks that if moral expertise is unattainable, then the state will always be badly run, whether it is governed by the poor or the rich, the many or the few. The defects of democracy will not be remedied merely by disenfranchising the many and installing a few wealthy aristocrats in office. What is needed to rule well is a high level of moral knowledge, and if the sort of knowledge Socrates is talking about is beyond our powers to acquire, then there would be no point in replacing a democracy with some other form of government.

When I speak of a 'moral expert' or of someone who has a 'high level of moral knowledge,' I mean someone who can satisfactorily defend an answer to the sorts of questions that are typically asked in the early dialogues: 'What is courage?,' 'What is justice?,' etc. Socrates thinks that an adequate answer to these questions will provide us with the standard we need in order to answer difficult and important practical questions[14]. Anyone who lacks such a standard – if he has any insight into himself – will be perplexed about a large number of urgent moral issues. Thus, our inability to define the virtues is a major moral deficiency. According to Socrates, since most people will never free themselves from false moral beliefs, they obviously will never become moral experts. True belief about certain matters – not to mention knowledge – is beyond them. If moral expertise is ever attained, it will be only by a few. But as yet, Socrates did not think that even a few had become moral experts[15]. And he was doubtful that any ever would.

Why does Socrates think that a move away from democracy would be worthwhile only on condition that moral experts take control? At first sight, this seems to be an irrational form of perfectionism. After all, if there are people who have freed themselves from *some* of the defects of the many – and Socrates thinks there are – then why shouldn't they rule, rather than the many? Why wouldn't Socrates have urged some of his followers to establish a new form of government – an oligarchy of Socratic believers? They would not be moral experts, since they lack definitions of the virtues, but they would all believe that injustice must never be done, under any circumstances. Wouldn't they therefore be better rulers than the many?

Let us recast this problem into the following terms: Suppose some friends and admirers of Socrates had asked him whether they should try to overthrow the democracy, and install themselves in power. They want him to say whether such a plan of action would be in their own best interest. I suggest that Socrates would give the following reply: 'If you want to pursue your own best interest, then you must try to become as virtuous as possible. And to do this, you must try to acquire knowledge of the virtues. But, the only way to acquire moral knowledge – though of course this is not a method that guarantees success or even progress – is to discuss moral questions every day, as I do. Now, if you enter a political career, you will have too little time, or none at all, to care for you own souls. If your revolutionary scheme succeeds, you will be responsible for ruling the state – a full-time job that will surely interfere with the more urgent task of acquiring moral knowledge.

Admittedly, as things now stand, we are being governed by men worse than ourselves. But the worst that could happen to us, as a result, is that we will become victims of injustice. And that is not a great evil: far worse to abandon our moral investigations, which give us the only chance we have to become fully virtuous men. So give up your political ambitions, turn yourselves into moral experts, and then – if you so choose – you can occupy yourselves with politics. Having completed the project of becoming a virtuous person, you can turn to the far less important task of protecting yourself from injustice. But until you become a moral expert, stay away from politics[16]'.

Suppose we press Socrates further, with the following question: 'Would the state be better ruled, if you and your followers were in office, and the democracy were dismantled? As you have just told us, you would never advise your friends or anyone else to become rulers, if they fell short of moral expertise. But if they turned down your advice and overthrew the democracy, would you say that Athens had thereby become a better governed city? After all, the individuals in power would believe that injustice must never be done. Doesn't that make them better rulers?'

I think Socrates would be ready with this reply: 'My followers', he would say, 'will *not* be better rulers than the many, if they lack definitions of the virtues. For without definitions, they will be unable to resolve the many difficult cases that will inevitably confront them. They will always want to do the right thing, but good intentions won't tell them which laws to adopt. They will refrain from all the acts that are obviously unjust to any sane adult, but that will not be enough to make their government significantly better than a democracy. And in difficult cases, they might lose their confidence that injustice is never to be done. So, when I tell my friends to stay out of politics, it's not because I don't care about the way Athens is governed. I care for my city even more than I cared for my parents[17], and so I want to see it competently managed. If I thought that Athens would be better governed by my followers than by the many, then I would be torn: I would have to admit that my city needs such men as my friends, but I would hate to advise them to pursue political careers. Fortunately, I am faced with no such dilemma. I think my followers should, for their own sakes, stay out of politics, but I don't think that this does any significant harm to my city.'

If Socrates would have replied in the way I suggest, then he was no active opponent of democracy, nor was he a partisan of the aristocratic political faction that existed in Athens[18]. He was a critic of democracy, and he thought that under proper conditions a democracy should be replaced by something better. But he thought that proper conditions would not exist until a few individuals had become moral experts.

4. Popper admits that Socrates was critical of the many, but insists that this is compatible with democratic values. Socrates 'felt that the way to improve the political life of the city was to educate the citizens to self-criticism'[19]. In other words, Popper's Socrates prefigured the liberalism of J.S. Mill: human beings are by and large equal in their capacities, and though the masses are now too fond of the lower pleasures, this is a condition that can be remedied through universal education. Thus Popper refers to Socrates' 'theory that moral excellence can be taught, and that it does not require any particular moral faculties, apart from the universal human intelligence'[20]. ... 'Socratic intellectualism is decidedly equalitarian. Socrates believed that everyone can be taught; in the *Meno*, we see him teaching a young slave ... in an attempt to prove that any uneducated slave has the capacity to grasp even abstract matters'[21]. Now, even though Socrates has reason to believe that virtue is teachable (since it is knowledge), that hardly shows that he thought that everyone or even a large majority are capable of learning it. After all, the skill of playing the flute can be taught – for there are expert teachers of it – but that does not mean that everyone can learn it, and in fact this ability may be quite rare. So Socrates can easily believe that if virtue is teachable, it will be learnable by only a few. (We will later see that Socrates took seriously the possibility that virtue may not be teachable at all: a point Popper overlooks.) Furthermore, nothing of moral significance is proved by the fact that in the *Meno* a slave is shown to have 'the capacity to grasp abstract matters.' True, a slave can learn geometry, but that hardly shows that he can acquire a knowledge of virtue, justice, the good, etc. None of the five passages we have looked at (in section 1) casts the slightest doubt on the capacity of the many to learn certain skills and certain subjects. Socrates assures us in the *Apology* that in their special spheres, the craftsmen do have the knowledge they claim[22]. It is knowledge of the most important matters – virtue and the good – that they lack, and as our five passages tell us, they will never acquire that knowledge. If egalitarianism means the belief in roughly equal human capacities to know the good and do what is right, then Socrates was no egalitarian. On the contrary, he believed that at most a few will ever

reach that level. And since most will never have knowledge of justice, they will never be deserving of political power. Popper makes an important concession when he says, 'It is not unlikely that he (Socrates) demanded ... that the best should rule, which would have meant, in his view, the wisest, or those who knew something about justice[23].' But since most people will never have knowledge about justice, they will never merit political office. So Popper ought to have conceded that, according to Socrates, democracy – the rule of the many – is a highly defective political form[24].

5. I said earlier that Socrates appreciated one of the great merits of Athenian democracy. What I have in mind will become clear after we examine another passage from the *Crito*. The Laws tell Socrates:

> ... You had seventy years in which you could have departed, if we did not satisfy you ... But you preferred neither Sparta nor Crete, which you frequently say are well-governed, nor any other Greek or foreign city, and you made fewer visits outside the city than do the lame, blind, or others who are crippled. In this way, it is clear that the city and we, the laws, have especially satisfied you, of all Athenians. For whom could the city satisfy, apart from its laws? (52e3-53a5)

Two schools of thought should be embarrassed by what the Laws say here. First, someone who thinks – as Popper does – that Socrates was an enemy of the closed society should be puzzled by the fact that he so frequently said that Sparta was well-governed. For fifth century Sparta was a closed society, if ever there was one. Spartan citizens were not allowed to live abroad, lest they be corrupted by foreign ideas; for the same reasons, visitors to Sparta were tightly controlled and periodically expelled. A Spartan boy, from the age of 7 onwards, would have been exposed to no ideas other than those sanctioned by the ministers of public education. They couldn't even enter the marketplace until they were 30[25]. How could Socrates – the great marketplace iconoclast and advocate of the examined life – have called such a society well-governed?

Popper's response to this problem is to throw out the line that creates our difficulty[26]. He suggests that the reference to Socrates' kind words for Sparta and Crete were not a genuine part of the original dialogue. Rather, it was added at some later time by an admirer of these cities – perhaps by Plato himself. But no editor has ever suspected this line on philological grounds, and it is of course a dangerous policy simply to throw out lines that conflict with one's preferred interpretation. I will try to show that Socrates' praise for Sparta can be explained.

There is another school of thought about Socrates that should be embarrassed by our passage: those who view him as an arch-enemy of Athenian democracy. For the *Crito* says that Socrates, more than any other citizen, is satisfied with the laws of Athens – so much so that he prefers his native city to Sparta and Crete. I take this to mean that although Socrates thinks that Sparta and Crete are well-governed, he nonetheless believes that Athens has better laws. But how so, if the laws are made by the many? Those who portray Socrates as an enemy of democracy give no explanation, partly because they have been led astray by Tredennick's mistranslation of the passage in question[27]. He has the Laws say that Sparta and Crete are Socrates' 'favourite models of good government'[28], though Socrates merely says that these cities are well-governed. Socrates' 'favourite model of good government,' among existing cities, is Athens, not Sparta or Crete.

6. So we are left with two problems: Why does Socrates so frequently say that Sparta is well-governed? And why is he more satisfied than any other citizen with the laws of Athens? I think we will be able to answer these questions if we put together a picture of how moral development takes place, according to the early dialogues. There seem to be four important stages:

(a) A child receives a conventional moral training from his parents. Admittedly, Socrates thinks that there are grave errors in conventional childhood education: for example, children are taught to help their friends and harm their enemies, to return injustice for injustice, etc., and these are opinions that Socrates attacks and finds impossible, in most cases, to eradicate. Nonetheless, he thinks that there are a great many truths in a traditional moral education as well, and he therefore believes that on balance a child who receives a conventional moral education is much better off than he would be without one. When Socrates challenges his fellow citizens to defend their views, he necessarily uses some of their true beliefs to get them to see that certain other beliefs are inconsistent with them. And in the *Crito* he agrees that children who have received a conventional moral training thereby owe their parents and the city a great debt of gratitude[29].

(b) Perhaps the best known fact about Socratic method is this: before someone can make progress in a subject and go beyond his naive stock of beliefs, he must be shocked out of his complacency. Those who have received a conventional moral training are unaware of the fact that they have a lot more progress to make before they can claim to be fully virtuous. And that is the point of Socratic questioning. By asking people for an account of the virtues, Socrates tries to get them to realize that they cannot do so, and that in fact they have inconsistent beliefs. So moral progress, according to Socrates, is not a matter of linear development. Some of the things we learn at the earliest stage of moral development have to be unlearned. Even though Socrates thinks that conventional moral education is highly valuable, the next stage of moral development involves the attempt to see the defects of that first stage.

(c) After someone has recognized his intellectual deficiencies, and has done a good deal of hard work, he might eventually reach the stage of moral development that Socrates himself has reached. This third stage is by no means the terminal point of moral growth; for Socrates still has a good many intellectual deficiencies, and he realizes this. He knows that some of his beliefs are inconsistent; he is caught in dilemmas to which he sees no easy solution[30]. Most important, he cannot give a satisfactory definition of the virtues, and he claims to have very little moral knowledge. Nonetheless, there is a good deal that he can do: he has systematically organized many of his moral beliefs, and has rejected a good deal that is false in a conventional moral education. There is no one who can defeat him in moral argument: even when he faces the leading intellectual figures of the day, his ethical theories come out unscathed, whereas theirs are shown to be riddled with contradictions[31].

(d) As I have said, there is a stage of moral development that Socrates thinks neither he nor anyone else has ever reached. In this sense, he would say that no one has yet become a fully virtuous person. Whoever does so would be able to give a satisfactory definition of the virtues, and with such a definition one would be able to decide how to act, even in the most difficult cases.

7. Let us now return to the first of the two problems we left hanging: why does Socrates think that Sparta and Crete are well-governed? My hypothesis is that he frequently calls them well-governed because they are so effective in moving their citizens up to stage (a) of moral development. In other words, Sparta and Crete went to greater lengths than other Greek cities to mould the individual's conception of right and wrong[32]; and since Socrates believes that conventional morality con-

tains a great many truths, he has good reason to use Sparta and Crete as models of well governed communities. If one important job of the city is to teach virtue, then the first stage of that job is to inculcate a large stock of true beliefs. That low-level process of education is precisely where Sparta and Crete excelled.

Recall now the second problem we left hanging: why was Socrates so satisfied with the legal system of Athens? My reply is that his native city provided the conditions in which citizens could move from stage (a) of moral development to stage (b).

Athens, like most other cities, encouraged the development of conventional beliefs about the virtues – the basic stock without which there could be no further progress. But it had an additional attraction in that it left its adult citizens free to make improvements in that ordinary moral training. Because Athenian democracy tolerated unorthodox speech and moral criticism, Socrates was at liberty to stand in the marketplace all day and challenge the conventional wisdom. As K.J. Dover has said recently, 'Tolerance of the free expression of intellectual criticism was at most times and in most circumstances a predominant characteristic of Athenian society'[33]. And of course, Socrates was quite conscious of this feature of Athenian law. He tells Polus, as he tries to restrain him from making a long speech rather than a short reply: 'You would be badly treated, my good man, if upon arriving in Athens, where there is more freedom to speak than anywhere in Greece, you alone should fail to get any'[34]. This freedom, of course, was not constitutionally guaranteed: there was no law in Athens that even remotely resembled the first amendment of the American constitution. The intellectual freedom enjoyed by the Athenians consisted in the absence of restrictive law, rather than in the guarantee of a right. That is, there were no laws in Athens, as there were in Sparta and Crete, that stifled the free flow of ideas. Athenian law permitted philosophical activity and moral criticism to a degree unparalleled by any other Greek city, and it surely was this 'negative' feature of the legal system that made Socrates one of the great admirers of his native city.

8. If the hypotheses I have put forward are correct, then Socrates had a firm understanding of the fact that legal systems can affect moral development. Popper was wrong when he said that Socrates was interested in people rather than institutions[35]. Equally misguided is the severe contrast Norman Gulley draws between the politics of Socrates and

Plato. He says that Plato 'rejects Socrates' belief that education is a personal affair, of individual by individual, and that education of this kind is the only proper education to promote the well-being of society. He (Plato) turns instead to state control of education'[36]. I suggest that Gulley's contrast is bogus: Plato never denies that it is individuals who teach individuals, nor does Socrates ever say that the state has no business controlling education. If my argument is correct, Socrates believes that a city's legal system has a great influence on the kind of education one person can give another. He presumably thinks that, ideally, the best legal system would be one that provided a setting in which a few individuals could reach stage (d) of moral development. Of course, no system of laws can by itself guarantee that citizens will reach that stage. But if there are moral experts, then a legal system can institutionalize their moral system and facilitate the transmission of that ethical code from one generation to another. The state, for example, could certify that certain individuals are moral experts, and it could require every citizen to learn as much as he could from such teachers. Gulley thinks that Socrates would have principled objections to such state interference with education. But the early dialogues tell a different story. Socrates thinks that the proper goal of a good ruler is to benefit the citizens[37], and that there is no greater benefit than moral development. If the stage could provide the right conditions for the teaching of virtue, Socrates would be delighted.

But can virtue be taught? Some scholars report that Socrates clearly answers in the affirmative[38], but I think the early dialogues tell a more complex story. In the *Protagoras*, for example, Socrates first argues that since no one – not even the wisest of Greeks – has been able to teach virtue, it must be unteachable[39]. And then, after arguing that virtue is knowledge, he admits that this conflicts with his earlier conclusion: for if virtue is knowledge, it must be teachable[40]. How are we to understand this contradiction? One possibility is that the earlier argument is entirely ironic: Plato intends us to reread the dialogue and to realize that the initial argument for the unteachability of virtue is intentionally fallacious. But I think there is good reason to reject this interpretation. For Socrates repeatedly says in the *Apology* that he does not teach virtue[41], though he wishes he could. With obvious irony, he says that he envies those sophists who can teach others to be good, for they have a superhuman form of wisdom, whereas he merely has human wisdom[42]. I take him to be suggesting that there is a great gulf between the wisdom of the gods and the wisdom of human beings

– a gulf that he strongly suspects is unbridgeable. He makes this same suggestion when he interprets the oracle in the following manner:

> It is likely, my friends, that the god is the one who is really wise, and what he means by his oracle is this: 'Human wisdom is worth little or nothing.' And it seems that in referring to Socrates, he is only using my name and making me an example, as though to say, 'Mankind, this one among you is wisest, who, like Socrates, has come to know that he is in truth worth nothing, in regard to wisdom.' (23a5–b4)

Socrates is saying that he sees little hope of ever moving from stage (c) of moral development to stage (d). His low level of wisdom has little chance of developing into the full moral understanding that the gods have. But he is not urging us to renounce our efforts to become as virtuous as possible. For there is always the possibility that we might discover a fully adequate moral theory, centred around definitions of the virtues. The only way we can tell what limits the gods have imposed on human nature is to try our utmost to acquire moral knowledge, and see how far we get. And if someone ever becomes a moral expert, he would, like any expert, be able to teach his subject to those few who have the ability to learn it.

The following analogy might help us understand what Socrates is saying: suppose a group of men is trying to shatter a certain metal. It is the sort of metal whose internal structure suggests that it should be breakable. But their best efforts over many years have been fruitless. Wouldn't it occur to them that perhaps the metal is unbreakable after all? If so, they would be in a dilemma comparable to the one Socrates describes in the *Protagoras*. He takes himself to be the wisest Athenian there is – for he alone is at stage (c) – but he is unable to teach virtue. If he could advance to stage (d), then he *would* be able to teach it. But despite his lifelong absorption in this project, a complete moral system eludes him, and in fact he takes himself to be quite distant from this goal ('he is in truth worth nothing, in regard to wisdom'). So he is strongly inclined to accept the hypothesis that even though virtue ought to be teachable – since it is knowledge – it is not teachable. By contrast, Protagoras thinks that the gods have decided to share their wisdom and virtue with all human beings[43]. The weakness of his theory emerges when he reveals how low his standard of virtue is: he counts anyone who has reached stage (a) as a virtuous man. Socrates' standard of virtue is far more demanding: when he asks whether virtue is teachable, he is asking whether an exact science of good and evil[44] can be imparted from one person to another. Obviously it cannot, if no one can ever acquire that high level of knowledge to begin with. But if someone ever does turn himself into a moral expert, then he will quickly provide a counterexample to the unteachability thesis. For those who have mastered a craft can teach it.

9. If my reading of the early dialogues is right, then there is a connection in Socrates' mind between the unteachability of virtue and the justification of democracy. Virtue is teachable if and only if there can be moral experts. And if there are no moral experts, then democracy is, for all its defects, no worse than any other political arrangement. Can we find any passage in the early dialogues that links democracy with unteachability in this way? The answer is yes. Here is what Socrates says in the *Protagoras* when he argues for the unteachability thesis:

> I call the Athenians wise, as do the other Greeks. For I see that whenever we meet in the Assembly, and need to take some common action about buildings, the builders are sent for as advisors, and similarly for shipbuilding and all other subjects they think are learnable and teachable. And if someone else whom they don't consider an expert (*dēmiourgon*) tries to advise them – even if he is quite handsome, rich and nobly born – they don't have a bit more tolerance for him. Rather, they laugh and shout until, having been shouted down, he gives up his attempt to speak, or the armed guards, at the request of the presiding officers, drag him away. That is how they treat a subject about which there is expertise (*en technēi*). But when they must make a decision about governing the city, then all alike stand up and give advice – joiner, smith, cobbler, merchant, shipowner, rich and poor, nobly or ill born. And no one rebukes them, as in the former case, on the grounds that he studied with no one and had no teacher, yet still tries to give advice. It's obvious that they don't think this is teachable (319b3 – d7).

Socrates wholeheartedly agrees with the idea of his fellow citizens that you should never seek the advice of an amateur when experts are at hand. And he also agrees that since moral experts do not exist, everyone should have his say in the Assembly. But notice that a highly undemocratic conclusion is lurking in this passage: if there were moral experts – i.e. people who know as much about virtue as other experts know about their own fields – then we should subordinate ourselves to them in the same way we look up to non-moral authorities. The *Crito* makes the same point in a passage we looked at earlier: Socrates tells his friend that in medical matters we rightly disregard the many and obey the expert; and he insists that this deference would be appropriate in the moral sphere as well – if there is someone who has the requisite knowledge[45]. Similarly, in the *Apology* Socrates lays down the principle that it is disadvantageous and shameful to disobey a superior, whether divine or human[46]. By a human superior, he cannot mean someone who occupies high office. For in the *Apology* he reminds his audience of his willingness to disobey that kind of superior: during the oligarchic revolution, he disobeyed the government in power, because their order to arrest Leon of Salamis was unjust[47]. The kind of superior Socrates must have in mind, when he says that we must always obey human superiors, is someone who is outstanding in virtue. I take him to be thinking of someone who has reached stage (d), for only such an individual would know enough to be an infallible guide for others. If anyone ever does reach that ultimate goal of moral development, then he is the one who should rule the state; he alone would be qualified for the job, and the rest of us would benefit from our obedience to him.

This interpretation of the early dialogues is at odds with one that Popper puts forward. He takes Socrates to believe that the role of an authority in moral matters is simply to get others to criticize themselves and to make up their own minds[48]. Norman Gulley follows Popper's lead, and says that according to Socrates 'the individual must be free to realize his own good'[49]. Similarly, Gregory Vlastos writes that for Socrates 'the role of the specialist and the expert should be only to offer guidance and criticism, to inform and clarify the judgement of the layman, leaving the final decision up to him'[50]. But none of these pronouncements tallies with the *Apology's* undeniable authoritarianism: if, as Socrates says, it is shameful ever to disobey a divine or human superior, then he is not bidding us merely to consult with moral authorities – if there ever are any – and then go off and make up our own minds. After all, Socrates would scoff at the proposal that we should give the commands of the gods a fair hearing, and then decide whether we want to obey them. The divine orders that come to us through dreams, prophecies and oracles[51] must be obeyed because to do otherwise would be shameful and contrary to our own interest. And the same deference should be paid to a human superior, if some human being ever becomes a moral expert and gives the rest of us commands. And of course, if a moral expert ever rules the state – something he can be expected to do, if he wants to avoid being ruled by worse men – then commands are what the other citizens can expect to receive.

Now, it is legitimate to raise the question of how someone who is not a moral expert can ever recognize that someone else is. I think that Socrates would reply that *he* at least could recognize moral expertise if he ever came across it: for he makes it his daily business to determine whether anyone else can define the virtues. To put the point more generally: if someone has reached a certain level of skill in a subject, then he thereby acquires the ability to determine whether certain others have far more skill. As we learn a subject, we see what the right questions are, and we take those who answer our questions in illuminating ways to be more advanced than we are. And of course, we frequently take others to be experts simply because we have confidence in the institutional framework that certifies that status. Given the proper social setting, certain individuals will come to be looked upon as moral experts; and as they teach others the ethical theory they have mastered, the confidence their students place in them will be confirmed.

At this point, someone might protest against my interpretation in the following way: 'you are completely forgetting about Socrates' most famous philosophical maxim: "the unexamined life is not worth living"[54]. Surely, if the rest of us just take orders from moral experts, and never think for ourselves, then we are not living an examined life. So even if there ever are moral experts, they would just ask moral questions, as Socrates does: they would try to get us to think for ourselves, but they wouldn't tell us what to believe or what to do. Such experts would therefore have no use for political power.' But this objection makes an important philosophical mistake, for there is no incompatibility between subordinating oneself to a person and trying to understand why the orders of that person are justified. To lead an examined life, one must subject one's moral beliefs to questioning; the false ones, if there are any, must be discovered and rejected, and the true ones must be systematically investigated until one sees why they are true. If you acquire your beliefs from someone you recognize as a moral expert, then you can rest assured that nothing you believe is false. But that still leaves a good deal of work for you to do. Just accepting what someone says doesn't give you much understanding of what he says. So I see no conflict between Socrates' statement that we would be wrong to disobey a moral superior and his statement that an unexamined life is not worth living. What Socrates must say is that if moral experts ever rule the state, they should not simply give commands to the rest of the citizens: they must also try to get the others to understand, so far as possible, the ethical theory that lies behind these commands. In such a society, all children could be taught the rudiments of the science of ethics, and as young adults, they might be challenged by licensed teachers to go beyond this elementary stage and acquire a deeper understanding of moral theory. Not many would master the complete theory of the virtues, but all could be challenged by experts to defend their beliefs in argument. Such a society, of course, is a dream world: Socrates believed it would never come to pass, and he wasted no effort – as Plato did – sketching out the institutions of his ideal society. This is not because he had no interest in politics or institutions, but because his first priority was to discover the ethical theory that would be taught by the expert rulers of such a utopia. When Socrates says that the many will never agree with him that injustice must never be done, his pessimism about changing the world is evident. Surely he realized that if the state controls education, as it did in Sparta, then the moral beliefs of the many could be manipulated in accordance with the wishes of the rulers. But he doubted that there would ever be leaders who would rule society in accordance with Socratic principles. He dissuaded his followers from pursuing a political career before they had mastered the correct ethical theory, and he had severe doubts about whether the complete moral system he sought would ever be discovered. That is why he thinks that the many will always be riddled with error[53].

I hope I have explained why it is inaccurate to portray Socrates either as a simple enemy or a simple friend of democracy. He thinks that the many will always rule badly, and he would prefer a society run by moral experts. But he holds out no hope for anything better than democracy, and he values the intellectual freedom provided by this political system. The authoritarianism we find in Plato's *Republic* has its roots in the early dialogues, but those who cherish free critical inquiry justifiably look upon Socrates as their patron saint. If I am right, these apparently conflicting elements in the early dialogues fit together to form a coherent political outlook[54].

Notes

1. Popper [1966] volume 1, p. 191.
2. Thus E. Barker [1959] p. 97; A.E. Taylor [1932] p. 96; A.D. Winspear and T. Silverberg [1939] p. 84; E. and N. Wood [1978] p. 97. W.K.C. Guthrie [1962] volume 3, pp. 415 – 6, quotes Barker and Taylor with approval, and rejects Popper.
3. I will pay almost no attention to Xenophon's portrait of Socrates, since I believe that wherever it differs from Plato's the latter is more reliable. For a brief defence of this assumption, see G. Vlastos: 'The Paradox of Socrates', in his [1971], pp. 1 – 3.
4. 31c4 – 32a2.
5. 31e1 – 4. All translations are my own.
6. 24c9 – 25a10.

7. 184c9 – d4.
8. 184e8 – 9.
9. 47a2 – 48a7.
10. 47c9 – d2.
11. 48b8 – 49e3.
12. 347b9 – d2.
13. Why was he so pessimistic? I suggest a simple explanation: he spent many years trying to change the beliefs of the many, but – with few exceptions – failed. Certainly he did not address his questions solely to a few wealthy admirers; rather, he tried to change the attitudes of everyone in the city. See 22c9 – e5, 29d5 – e3, 30e1 – 31a1, 33a6 – b3. These last remarks might be dismissed as Plato's attempt to cover up the fact that Socrates was concerned exclusively with a few aristocratic youths. But I would be sceptical of any attempts to read the *Apology* as a whitewash. Such passages as (a) and (b), cited in section 1 above, refute this hypothesis. Some scholars take Socrates to believe that the many are beyond improvement because they engage in physical labour. Thus Guthrie [1962] volume 3, pp. 410 – 11. But the only evidence for this comes from Xenophon (e.g. *Memorabilia* III vii 6); no such suggestion is made in Plato's early dialogues.
14. *Euthyphro* 6e3 – 6.
15. I assume throughout this paper that Socrates has no answer to the definitional questions he raises. This assumption could be challenged, of course, but I cannot discuss the issue more fully here. For the connection between being an expert and having a definition, see *Laches* 184e11 – 186e3, 189d4 – 190c2.
16. For Socrates' belief that it is far more important to avoid acting unjustly than to avoid suffering injustice, see *Gorgias* 469b8 – c2, 477b1 – 479a3, 508a8 – c5, 509c6 – 7, 512a2 – b2. At *ibid.* 509d7 – e2, he says that one cannot avoid acting unjustly until one has acquired a certain 'power or craft', i.e. until one has the knowledge of virtue that comes with definitions. At *ibid.* 514a5 – 514b4, he says that one should refrain from putting oneself forward as a public servant until one has perfected one's abilities in private life. This point is repeated at the close of the dialogue (527d2 – 5).
17. *Crito* 51a7 – b4.
18. Here I am disagreeing with the views found in Winspear and Silverberg [1939], and in E. and N. Wood [1978].
19. Popper [1966] volume 1, p. 130.
20. *Ibid.* p. 128.
21. *Ibid.* p. 129.
22. 22c9 – e5.
23. Popper [1966] volume 1, p. 128.
24. Popper also appeals to *Gorgias* 488b8 – 489c7 to support his thesis that Socrates was an egalitarian. See his [1966] volume 1, pp. 91, 117, 128, 254. Against this, see Irwin [1977]. I side with Irwin.
25. On Sparta, see Plutarch's *Life of Lycurgus*, especially 9, 15 – 17, 24 – 27; Plato's *Hippias Major* 283b4 – 286c2; Xenophon's *Constitution of the Lacedaimonians* 3 – 6; Demosthenes *Against Leptines* 106. On Crete, see Plato's *Laws* 634d7 – e4.
26. Popper [1966] volume 1, p. 305.
27. *The Last Days of Socrates*, reprinted in E. Hamilton and H. Cairns (*eds*) [1963]. This is the translation used by E. and N. Wood [1978] p. 99.
28. 52e5 – 6.
29. 50d1 – 51c4.
30. *Gorgias* 527d5 – e1; *Hippias Minor* 372d7 – e1, 376b8 – c6; *Protagoras* 361a2 – c6. This feature of Socrates is frequently overlooked.
31. *Gorgias* 508e6 – 509a7.
32. Aristotle praises Sparta and Crete for these same reasons. See *Nicomachean Ethics* 1102a8 – 11, 1180a24 – 28. But this does not prevent him from criticizing Sparta on other grounds. See *Eudemian Ethics* 1248b38 – 40, *Politics* 1271b7 – 10. Fifth century Spartans were widely reputed to be exceedingly law-abiding. See Xenophon, *Constitution of the Lacedaimonians* 8, and Herodotus *Histories* VII 104.
33. Dover [1976] p. 54.

34. *Gorgias* 461e1 – 3.
35. Popper [1966] volume 1, p. 191.
36. Gulley [1968] p. 176.
37. *Republic* 345e2 – 347a6.
38. See for example, T. Irwin [1977] p. 102.
39. 319a10 – 320c1.
40. 361a1 – b3.
41. 19d8 – e1, 20d9 – e2, 33a5 – 6, 33b5 – 6.
42. 19d8 – 20e3.
43. 322c1 – d5.
44. 356e8 – 357b6.
45. 47c8 – d3.
46. 29b6 – 7.
47. 32c4 – e1.
48. Popper [1966] volume 1, pp. 129 – 130.
49. Gulley [1968] p. 177.
50. Vlastos [1971] p. 20.
51. *Apology* 33c5 – 6.
52. *Ibid.* 38a5 – 6.
53. Objection: 'If there ever are moral experts who rule the state in accordance with a new educational programme, then perhaps they will be able to teach *everyone* else to become moral experts. You have no evidence that Socrates wanted to exclude this possibility.' I think that passages (a) and (b), cited in section 1, provide such evidence. Socrates' contrast between the one who is wise and the ignorant many is not a contrast between someone who exists and the many. Rather, he is saying that at most no more than a few could ever become moral experts. He evidently thinks that the science of good and evil is so difficult that at best a handful could ever master it.
54. The material in this paper is drawn from chapter VII of my book, *Socrates and the State*, Princeton, N.J.: Princeton University Press, 1984.

BIBLIOGRAPHY

Ackermann, R.J. [1976]: *The Philosophy of Karl Popper*. Amherst: University of Massachusetts Press.

Adorno, T., Albert, H., Habermas, J., Popper, K.R. *et al.* [1976]: *The Positivist Dispute in German Sociology*. New York: Harper and Row. English translation of *Der Positivismusstreit in der deutschen Soziologie*. Neuwied u. Berlin: Luchterhand, 1969.

Adorno, T. and Horkheimer, M. [1972]: *Dialectic of Enlightenment*. New York: Herder and Herder.

Agassi, J. [1977]: *Towards a Rational Philosophical Anthropology*. The Hague: Martinus Nijhoff.

Albert, H. [1964]: 'Social Science and Moral Philosophy' in M. Bunge (*ed.*): *The Critical Approach to Science and Philosophy*, pp. 385 – 409. Glencoe: The Free Press.

Albrecht, R. [1973]: *Sozialtechnologie und ganzheitliche Sozialphilosophie*. Bonn: Bouvier Verlag Herbert Grundmann.

Alston, W.P. [1964]: 'The Elucidation of Religious Statements', in W.R. Reese and E. Freeman (*eds.*): *Process and Divinity: The Hartshorne Festschrift*. La Salle: Open Court.

Althusser, L. [1971]: *Lenin and Philosophy and Other Essays*. London: New Left Books.

Apel, K-O. [1980]: *Towards a Transformation of Philosophy*. London: Routledge and Kegan Paul. English translation of parts of *Transformation der Philosophie*, volumes 1 and 2. Frankfurt am Main: Suhrkamp, 1972 – 1973.

Arens, W. [1979]: *The Man-Eating Myth*. London: Oxford University Press.

Aristotle: *The Works of Aristotle Translated into English*. W.D. Ross (*ed.*) Oxford: Clarendon, 1908 – 1952.

Arlow, J. [1960]: 'Psychoanalysis and Scientific Method', in S. Hook (*ed.*): *Psychoanalysis, Scientific Method and Philosophy*. New York: Grove Press.

Averill, E. and Keating, B.F. [1981]: 'Does Interactionism Violate a Law of Classical Physics?', *Mind* 90: 102 – 107.

Ayer, A.J. [1954]: *Philosophical Essays*. London: Macmillan.

Bagish, H.H. [1981]: 'Confessions of a former Cultural Relativist', in E. Angeloni (*ed.*): *Anthropology 82 – 83 – Annual Editions*. Guildfort, CT: Dushkin Publishing Group.

Baier, H. v. [1966]: 'Soziologie und Geschichte', *Archiv für Rechts und Sozialphilosophie* 52: 67 – 91.

Barker, E. [1959]: *Political Thought of Plato and Aristotle*. New York: Dover Publications.

Bayertz, K. and Schleifstein, J. [1977]: *Mythologie der 'Krischen Vernunft'*. Köln: Pahl-Rugenstein Verlag.

Beck, H. [1974]: 'Der Positivismusstreit', in H. Rombach (*ed.*), *Wissenschaftstheorie* 1. Freiburg: Herder.

Bettelheim, C. [1978]: *Class Struggles in the USSR. Second Period: 1923 – 1930*. New York and London: Monthly Review Press.

Bhaskar, R. [1978]: *A Realist Theory of Science*. Second Edition. Sussex: Harvester.

Blackburn, S. [1971]: 'Moral Realism', in J. Casey (*ed.*): *Morality and Moral Reasoning*. London: Methuen.

Blackburn, S. [1981]: 'Reply: Rule-Following and Moral Realism' in S.H. Holtzman (*ed.*): *Wittgenstein: To Follow a Rule*. London: Routledge and Kegan Paul.
Bloor, D. [1978]: *Knowledge and Social Imagery*. London: Routledge and Kegan Paul.
Boyle, E. [1974]: 'Karl Popper's *Open Society*: A Personal Appreciation' in P.A. Schilpp (*ed.*): *The Philosophy of Karl Popper*. La Salle: Open Court.
Brill, A. [1948a]: *Basic Principles of Psychoanalysis*. New York: Washington Square Press.
Brill, A. [1948b]: *Lectures on Psychoanalytic Psychiatry*. New York: Vintage Books.
Bubner, R. [1981]: *Modern Germany Philosophy*. Cambridge University Press.
Cajori, F. [1962]: 'Appendix' in F. Cajori (*ed.*), *Sir Isaac Newton's Mathematical Principles of Natural Philosophy and his System of the World*. Berkeley: University of California Press.
Carter, C. [1954]: *An Encyclopedia of Psychoanalytic Astrology*. London: The Theosophical Publishing House.
Churchland, P. [1970]: 'The Logical Character of Action Explanations', *Philosophical Review* 79: 214 – 236.
Cioffi, F. [1970]: 'Freud and the Idea of a Pseudo-Science', and 'Reply', in R. Borger and F. Cioffi (*eds.*): *Explanation in the Behavioural Sciences*. Cambridge University Press.
Cioffi, F. [1976]: 'Was Freud a Liar?', *Orthomolecular Psychiatry* 5: 275 – 80.
Claudin, F. [1980]: *Marx, Engels et la Révolution de 1848*. Paris: Maspero.
Cohen, L.J. [1980]: 'The Problem of Natural Laws', in D.H. Mellor (*ed.*), *Prospects for Pragmatism*. Cambridge University Press.
Cornforth, M. [1968]: *The Open Philosophy and the Open Society: A Reply to Dr. Karl Popper's Refutations of Marxism*. London: Lawrence and Wishart.
Cowling, M. [1962]: *Mill and Liberalism*. Cambridge University Press.
Currie, G.P. [1978]: 'Popper's Evolutionary Epistemology: a Critique'. *Synthese* 37: 413 – 431.
Dalbiez, R. [1941]: *Psychoanalytic Method and the Doctrine of Freud*. London: Longman, Green & Co.
Davidson, D. [1980]: *Essays on Actions and Events*. Oxford: Clarendon Press.
Demosthenes: *The Oration against Leptines*. J.R. King (*ed.*) London: Macmillan, 1881.
Dennett, D. [1978]: *Brainstorms*. Hassocks, Sussex: Harvester Press.
Dennett, D. [1981]: 'Three Kinds of Intentional Psychology', in R. Healey (*ed.*) *Reduction, Time and Reality*. Cambridge University Press.
Donagan, A. [1974]: 'Popper's Examination of Historicism' in P.A. Schilpp (*ed.*): *The Philosophy of Karl Popper*: 905 – 924. La Salle: Open Court.
Dore, R.P. [1961]: 'Function and Cause', *American Sociological Review* 26: 843 – 53.
Dover, K.J. [1976]: 'The Freedom of the Intellectual in Greek Society', *Talanta*, 17: *Proceedings of the Dutch Archaeological and Historical Society*.
Duhem, P. [1968]: *The Aim and Structure of Physical Theory*. New York: Atheneum.
Dworkin, R.M. [1978]: 'Liberalism', in S. Hampshire (*ed.*): *Public and Private Marality*. Cambridge University Press.
Eddy, M.B. [1875]: *Science and Health*. Boston: Allison V. Stewart.
Ellis, B. [1979]: *Rational Belief Systems*. Oxford: Basil Blackwell.
Escalona, S. [1952]: 'Problems in Psychoanalytic Research', *International Journal of Psychoanalysis* 33: 11 – 21.
Ferber, C. v. [1976]: 'Der Werturteilsstreit 1909/1959. Versuch einer wissenschaftsgeschichtlichen Interpretation', in E. Topitsch (*ed.*): *Logik der Sozialwissenschaften*, 165 – 180. Köln: Kiepenheuer und Witsch.
Foot, P. [1978]: *Virtues and Vices and other essays in Moral Philosophy*. Oxford: Basil Blackwell.
Frank, J. [1961]: *Persuasion and Healing*. London and Baltimore: Johns Hopkins Press.
Freeman, M. [1975]: 'Sociology and Utopia: Some Reflections on the Social Philosophy of Karl Popper', *The British Journal of Sociology* 26: 20 – 34.
Freud, S. and Breuer, J. [1895]: *Studies in Hysteria*. New York, 1966.
Freud, S. [1896]: 'The Aetiology of Hysteria', in *Collected Papers*, volume 1. London: The Hogarth Press and the Institute of Psycho-analysis, 1924.
Freud, S. [1898]: 'Sexuality in the Aetiology of the Neuroses', in *Collected Papers*, volume 1.
Freud, S. [1900]: *The Interpretation of Dreams*. London: Allen & Unwin, 1954.
Freud, S. [1904a]: 'On Psychotherapy', in *Collected Papers*, volume 1.
Freud, S. [1904b]: 'Freud's Psychoanalytic Method', in *Collected Papers*, volume 1.

Freud, S. [1905a]: 'My views on the Part played by Sexuality in the Aetiology of the Neuroses', *Collected Papers*, volume 1.

Freud, S. [1905b]: *Three Essays on the Theory of Sexuality.* New York: Basic Books, 1963.

Freud, S. [1909a]: 'Notes upon a case of Obsessional Neurosis', *Standard Edition* volume 10. London: The Hogarth Press, 1953 – 66.

Freud, S. [1909b]: 'Bemerkungen über einen Fall von Zwangneurose', *Gesammelte Werke*, Band 7. Frankfurt am Main, 1972.

Freud, S. [1910]: 'The Future of Psychoanalytic Psychotherapy', *Collected Papers*, volume 2.

Freud, S. [1916]: *Introductory Lectures on Psycho-analysis.* London: Hogarth Press, 1963.

Freud, S. [1923]: 'Remarks on the Theory and Practice of Dream Interpretation', in *Collected Papers*, volume 5.

Freud, S. [1924]: 'The Question of Lay Analysis', in *Two Short Accounts of Psychoanalysis.* London: Pelican, 1962.

Freud, S. [1925]: *An Autobiographical Study.* London: Hogarth Press, 1950.

Freud, S. [1933]: *New Introductory Lectures.* London. Hogarth Press, 1949.

Gazzaniga, M.S. [1970]: *The Bisected Brain.* New York: Appleton-Century-Crofts.

Geyl, P. [1955]: *Debates with Historians.* Groningen: Wolters.

Giddens, A. [1974]: *Positivism and Sociology.* London: Heinemann.

Glover, E. [1956]: *Freud or Jung.* New York: W.W. Norton.

Glymour, C. [1974]: 'Freud, Kepler and the Clinical Evidence', in R. Wollheim (*ed.*): *Freud: a Collection of Critical Essays.* New York: Anchor Books.

Godelier, M. [1978]: 'Infrastructures, Societies and History', *New Left Review* 112: Nov. – Dec., 84 – 96.

Gramsci, A. [1971]: *Selections from the Prison Notebooks.* London: Lawrence and Wishart.

Gramsci, A. [1977]: *Selections from Political Writings (1910 – 1920).* London: Lawrence and Wishart.

Grünbaum, A. [1976]: 'Is Falsifiability the Touchstone of Scientific Rationality?', in R.S. Cohen *et. al.* (*eds.*): *Boston Studies in the Philosophy of Science,* volume 39. Dordrecht: D. Reidel.

Grünbaum, A. [1977]: 'How Scientific is Psychoanalysis?', in R. Stern *et al.* (*eds..*): *Science and Psychotherapy.* New York: Haven Publishing.

Grünbaum, A. [1979a]: 'Epistemological Liabilities of the Clinical Appraisal of Psychoanalytic Theory', *Psychoanalysis and Contemporary Thought*, volume 2, number 2.

Grünbaum, A. [1979b]: 'Is Freudian Psychoanalytic Theory Pseudo-Scientific by Karl Popper's Criterion of Demarcation', *American Philosophical Quarterly* 16: 131 – 41.

Gulley, N. [1968]: *The Philosophy of Socrates.* London: Macmillan.

Guthrie, W.K.C. [1962]: *History of Greek Philosophy.* Cambridge University Press.

Habermas, J. [1970]: 'Towards a Theory of Communicative Competence', *Inquiry* 13: 360 – 375.

Habermas, J. [1973]: *Theory and Practice.* Translated by J. Viertal. Boston: Beacon Press.

Hamilton, E. and Cairns, H. (*eds.*) [1963]: *The Collected Dialogues of Plato.* Princeton University Press.

Hanson, N.R. [1972]: 'Quantum Mechanics, Philosophical Implications of', in P. Edwards (*ed.*): *The Encyclopedia of Philosophy* 7. New York: Macmillan.

Hare, R.M. [1952]: *The Language of Morals.* Oxford: Clarendon Press.

Hare, R.M. [1963]: *Freedom and Reason.* Oxford: Clarendon Press.

Harris, M. [1968]: *The Rise of Anthropological Theory.* New York: Thomas Y. Crowell.

Harris, M. [1979]: *Cultural Materialism: the Struggle for a Science of Culture.* New York: Random House.

Hart, H.L.A. [1961]: *The Concept of Law.* Oxford: Clarendon Press.

Heller, A. [1978]: 'The Positivism Dispute as a Turning Point in German Post-War Theory', *New German Critique* 15: 49 – 56.

Herodotus: *The Histories*, translated by A. de Selincourt. Harmondsworth: Penguin Books, 1954.

Howson, C. [1973]: 'Must the Logical Probability of Laws be Zero?', *British Journal for the Philosophy of Science* 24: 152 – 182.

Huff, T.E. [1981]: 'On the Methodology of the Social Sciences: A Review Essay', Part I, *Philosophy of the Social Sciences* 11: 461 – 475.

Iggers, G.C. [1968]: *The German Conception of History.* Middleton: Wesleyan University Press.

Irwin, T. [1977]: *Plato's Moral Theory*. Oxford: Clarendon.
Jacoby, R. [1981]: *Dialectic of Defeat*. Cambridge University Press.
Jarvie, I.C. [1964]: *The Revolution in Anthropology*. London: Routledge and Kegan Paul.
Jarvie, I.C. [1972]: *Concepts and Society*. London: Routledge and Kegan Paul.
Johansson, I. [1975]: *A Critique of Karl Popper's Methodology*. Stockholm: Akademiforlaget.
Jones, E. [1923]: 'War Shock and Freud's Theory of the Neuroses', in *Papers on Psychoanalysis* London: Balliere, Tindall and Cox, 1950.
Jones, E. [1939]: 'Sigmund Freud', *Proceedings of the Royal Society of Medicine* 33: 111 – 15.
Kaplan, D. and Manners, R.A. [1972]: *Culture Theory*. Englewood Cliffs, N.J.: Prentice-Hall.
König, R. [1959]: 'Germany' in J. Roucek (*ed.*): *Contemporary Sociology*. London: Peter Owen.
Koertge, N. [1974]: 'On Popper's Philosophy of Social Science' in K.F. Schaffner and R.S. Cohen (*eds.*): *Boston Studies in Philosophy of Science*, volume 20. Dordrecht: Reidel.
Koertge, N. [1975]: 'Popper's Metaphysical Research Program for the Human Sciences', *Inquiry* 18: 437 – 462.
Korsch, K. [1970]: *Marxism and Philosophy*. London: New Left Books.
Krige, J. [1978]: 'Popper's Epistemology and the Autonomy of Science', *Social Studies of Science* 8: 287 – 307.
Kuhn, T.S. [1962]: *The Structure of Scientific Revolutions*. University of Chicago Press.
Lakatos, I. [1978]: 'Newton's Effect on Scientific Standards' in J. Worrall and G. Currie (*eds.*): *Imre Lakatos: Philosophical Papers*, volume 1: The Methodology of Scientific Research Programmes, pp. 193 – 222. Cambridge University Press.
Lakatos, I. and Musgrave, A. (*eds.*) [1968]: *Problems in the Philosophy of Science*. Amsterdam: North Holland.
Lee, D.E. and Beck, R.N. [1954]: 'The Meaning of "Historicism"', *American History Review* 57: 568 – 577.
Lorenzen, P. [1970]: 'Szientismus versus Dialektik' in R. Bubner, C. Cramer and R. Wiehl (*eds.*): *Hermeneutik und Dialektik*, pp. 57 – 72. Tubingen: J.C.B. Mohr.
Lührs, G., Sarrazin, T., Spreer, F., and Tietzel, M. (*eds.*) [1975]: *Kritischer Rationalismus und Sozialdemokratie*. Berlin: Verlag J.H.W. Dietz.
Lukes, S. [1972]: 'Methodological Individualism Reconsidered' in D. Emmet and A. Macintyre (*eds.*), *Sociological Theory and Philosophical Analysis*, pp. 76 – 88. London: Macmillan.
Mackie, J.L. [1977]: *Ethics: Inventing Right and Wrong*. Harmondsworth, Middlesex: Penguin Books.
Magee, B. *Popper*. London: Fontana.
Marcuse, H. [1972]: 'Karl Popper and the Problem of Historical Laws' in H. Marcuse: *Studies in Critical Philosophy*, pp. 191 – 208. Boston: Beacon Press.
Marx, K. [1953]: *Grundrisse der Kritik der politischen Oekonomie (Rohentwurf) 1857 – 1858*. Berlin: Dietz.
Marx, K. [1970]: *A Contribution to the Critique of Political Economy*. Moscow: Progress Publishers.
Marx, K. [1973]: *Grundrisse. Foundations of the Critique of Political Economy*. Penguin: New Left Books.
Marx, K. [1975]: *Texts on Method*, T. Carver (ed.). Oxford: Blackwell.
Marx, K. [1976 – 81]: *Capital. A Critique of Political Economy*. (Three volumes). Harmondsworth: Penguin: New Left Review. Vol. 1, 1976; Vol. 2, 1978; Vol. 3, 1981.
Marx, K. and Engels, F. [1953]: *Selected Correspondence*. Moscow: Foreign Language Publishing House.
Marx, K. and Engels, F. [1956]: *Werke*. Berlin: Dietz.
Marx, K. and Engels, F. [1970]: *Selected Works in Three Volumes* Moscow: Progress Publishers.
Marx, K. and Engels, F. [1975]: *Collected Works*. Moscow: Progress Publishers.
Maxwell, G. [1976]: 'The Role of Scientific Results in Theories of Mind and Brain' in G.G. Globus, G. Maxwell and I. Savodnik (*eds.*): *Consciousness and the Brain: A Scientific and Philosophical Enquiry*, pp. 317 – 328. New York and London: Plenum Press.
Mepham, J. [1979]: 'From the Grundrisse to Capital: the Making of Marx's Method' in J. Mepham and D. Hillel-Ruben (*eds.*): *Issues in Marxist Philosophy* volume 1. Sussex: Harvester.
Miles, I. [1975]: *The Poverty of Prediction*. Lexington: D.C. Heath.
Mill, James. [1975]: *Essay on Government*. Indianapolis: Bobbs-Merrill.

Mill, J.S. [1914]: *Utilitarianism, Liberty, Representative Government*. London: J.M. Dent and Sons.
Moore, G.E. [1903]: *Principia Ethica*. Reprinted, 1929. Cambridge University Press.
Moore, G.E. [1909 – 10]: 'The Subject Matter of Psychology', *Proceedings of the Aristotelian Society* 10: 36 – 62.
Mountcastle, V.B. [1975]: 'The View from Within: Pathways to the Study of Perception', *The Johns Hopkins Medical Journal*, 136: 109 – 131.
Nassen, U. (*ed.*) [1979]: *Studien zur Entwicklung einer materialen Hermeneutik*. Munich: Wilhelm Fink.
O'Hear, A. [1980]: *Karl Popper*. London: Routledge and Kegan Paul.
Papineau, D. [1978]: *For Science in the Social Sciences*. London: Macmillan.
Parekh, B.C. [1982]: *Contemporary Political Thinkers*. Oxford: Martin Robinson.
Plato: *Platonis Opera*. J. Burnet (*ed.*). Oxford: Clarendon Press, 1905 – 1946.
Plutarch: 'Life of Lycurgus' in *Plutarch's Lives*, translated by J. Langhorne and W. Langhorne. London: Walter Scott, 1886.
Popper, K.R. [1948]: 'Prediction and Prophecy and their Significance for Social Theory' in E.W. Beth *et.al.*(*eds.*): *Library of the Tenth International Congress of Philosophy*, volume 1: Proceedings of the Tenth International Congress of Philosophy. Amsterdam: North-Holland. Republished as chapter 16 in Popper [1963], pp. 336 – 346.
Popper, K.R. [1959]: *The Logic of Scientific Discovery*. London: Hutchinson.
Popper, K.R. [1960]: *The Poverty of Historicism*. Second edition (revised). London: Routledge and Kegan Paul.
Popper, K.R. [1961]: *The Poverty of Historicism*. Reprint of second edition (with some corrections). London: Routledge and Kegan Paul.
Popper, K.R. [1962]: *The Open Society and its Enemies*. Fourth Edition (revised). London: Routledge and Kegan Paul.
Popper, K.R. [1963]: *Conjectures and Refutations*. London: Routledge and Kegan Paul.
Popper, K.R. [1964]: *The Poverty of Historicism*. Reprint of 1961 edition. New York and Evanston: Harper and Row.
Popper, K.R. [1966]: *The Open Society and Its Enemies*. Fifth Edition (revised). London: Routledge and Kegan Paul.
Popper, K.R. [1967]: 'Le Rationalité et le Statut de Principe de Rationalité', in E.M. Claassen (*ed.*): *Les Fondements Philosophiques des Systemes Economiques*, pp. 142 – 150. Paris: Payot.
Popper, K.R. [1968a]: *The Logic of Scientific Discovery*. Fifth edition (revised). London: Hutchinson.
Popper, K.R. [1968b]: 'Remarks on the Problems of Demarcation and of Rationality', in I. Lakatos and A. Musgrave (*eds.*): *Problems in the Philosophy of Science*, pp. 88 – 102. Amsterdam: North-Holland.
Popper, K.R. [1969]: *Conjectures and Refutations*. Third edition (revised). London: Routledge and Kegan Paul.
Popper, K.R. [1970]: 'Normal Science and Its Dangers', in I. Lakatos and A. Musgrave (*eds.*): *Criticism and the Growth of Knowledge*. Cambridge University Press.
Popper, K.R. [1972]: *Objective Knowledge: An Evolutionary Approach*. Oxford: Clarendon Press.
Popper, K.R. [1974]: 'Replies to my Critics', in P.A. Schilpp (*ed.*): *The Philosophy of Karl Popper*, Book II. La Salle: Open Court.
Popper, K.R. [1976]: *Unended Quest*. La Salle: Open Court.
Popper, K.R. [1979]: *Objective Knowledge: An Evolutionary Approach*. Revised edition. Oxford: Clarendon Press.
Popper, K.R. [1980]: 'Three Worlds', in S. McMurrin (*ed.*): *The Tanner Lectures on Human Values*. Salt Lake City: University of Utah Press.
Popper, K.R. and Eccles, J.C. [1977]: *The Self and Its Brain*. Berlin: Springer Verlag.
Pucetti, R.P. [1974]: Review of *The Self and Its Brain, Philosophical Books* 29: 87 – 90.
Pucetti, R.P. [1978]: 'Unravelling the World Knot', *British Journal for the Philosophy of Science* 29: 143 – 148.
Pucetti, R.P. and Dykes, R.W. [1978]: 'Sensory Cortex and the Mind-Brain Problem', *The Behavioral and Brain Sciences* 3: 337 – 375.

Rank, O. [1927]: Review of *Inhibitions, Symptoms and Anxiety, Mental Hygiene* 11: 181 – 8.

Ravetz, J. [1981]: 'Ideologische Ueberzeugungen in der Wissenschaftstheorie' in *Versuchungen Aufsatze zur Philosophie Paul Feyerabends*, volume 1. Frankfurt a.M.: Suhrkamp.

Raymond, P. [1982]: *Le Resistible Fatalité de l'Histoire*. Paris: J.E. Hollier/Albin Michel.

Reiss, H. (*ed.*) [1970]: *Kant's Political Writings*. Cambridge University Press.

Robbins, L. [1935]: *An Essay on the Nature and Significance of Economic Science*. Second edition. London: Macmillan.

Robbins, L. [1938]: 'Live and Dead Issues in the Methodology of Economics', *Economica* N.S. 5: 342 – 352.

Rorty, R. [1979]: *Philosophy and the Mirror of Nature*. Princeton University Press.

Russell, B. [1948]: *Human Knowledge: Its Scope and Limits*. New York: Simon and Schuster.

Russell, B. [1959]: *My Philosophical Development*. New York: Simon and Schuster.

Schilpp, P.A. (*ed.*) [1974]: *The Philosophy of Karl Popper*. Library of Living Philosophers volume 14. La Salle: Open Court.

Schmidt, P.F. [1968]: 'Some Criticisms of Cultural Relativism', in R.A. Manners and D. Kaplan (*eds.*): *Theory in Anthropology: a Source book*, pp. 169 – 174. Chicago: Aldire Publishing Co.

Schupp, F. [1975]: *Popper's Methodologie der Geschichtswissenschaft*. Bonn: Bouvier Verlag Herbert Grundmann.

Seiffert, H. [1975]: *Einführung in die Wissenschaftstheorie*. Munich: Verlag C.H. Beck.

Skagestad, P. [1975]: *Making Sense of History: The Philosophies of Popper and Collingwood*. Oslo: Universitetsforlaget.

Smart, J.J.C. [1959]: 'Sensations and Brain Processes', *The Philosophical Review* 68: 141 – 156.

Spinner, H.F. [1978]: *Popper und die Politik. Volume 1, Geschlossenheitsprobleme*. Berlin: Verlag J.H.W. Dietz.

Stevenson, C.L. [1937]: 'The Emotive Meaning of Ethical Terms' *Mind* 46: 14 – 31.

Strawson, P.F. [1959]: *Individuals*. London: Methuen.

Suchting, W. [1972]: 'Marx, Popper and "Historicism"', *Inquiry* 15: 235 – 66.

Suchting, W. [1983]: *Marx. An Introduction*. Sussex: Harvester.

Taylor, A.E. [1932]: *Socrates*. London: Peter Davies.

Thiel, C. [1972]: *Grundlagenkrise und Grundlagenstreit*. Meisenheim am Glan: Verlag Anton Hain.

Topitsch, E. (*ed.*) [1976]: *Logik der Sozialwissenschaften*. Köln: Kiepenheuer und Witsch.

Urbach, P. [1978a]: 'Is Any of Popper's Arguments Against Historicism Valid?', *British Journal for the Philosophy of Science* 29: 117 – 130.

Urbach, P. [1978b]: 'The Objective Promise of a Research Programme', in G. Radnitzky and G. Andersson (*eds.*), *Progress and Rationality in Science*. Dordrecht: Reidel.

Urbach, P. [1981]: 'On the Utility of Repeating the "Same" Experiment', *Australian Journal of Philosophy* 59: 151 – 162.

Vlastos, G. [1971]: *The Philosophy of Socrates*. New York: Anchor Books.

von Hayek, F.A. [1944]: *The Road to Serfdom*. London: G. Routledge and Sons.

Wellmer, A. [1967]: *Methodologie als Erkenntnistheorie*. Frankfurt: Suhrkamp.

Wellmer, A. [1974]: *Critical Theory of Society*. New York: The Seabury Press. English translation of *Kritische Gesellschaftstheorie und Positivismus*. Frankfurt am Main: Suhrkamp, 1969.

Wilkins, B.T. [1978]: *Has History Any Meaning?* Hassocks, Sussex: Harvester Press.

Winch, P. [1972]: *Ethics and Action*. London: Routledge and Kegan Paul.

Winspear, A.D. and Silverberg, T. [1939]: *Who was Socrates?* New York: The Cordon Company.

Witschel, G. [1971]: *Wertvorstellung im Werk Karl R. Poppers*. Bonn: Bouvier Verlag Herbert Grundmann.

Wood, E.M. and Wood, N. [1978]: *Class Ideology and Ancient Political Theory*. Oxford: Blackwell.

Woodworth, R.S. [1913]: *Review of Nervous and Mental Disease Monograph Series*, volume 38, number 291, December 26.

Woolf, H.H. [1968]: 'The Psychosomatic Approach to Psychological Disorders: Contributions and Limitations of Psychoanalysis', *British Journal of Medical Psychology* 41: 343 – 8.

Wright, L. [1976]: *Teleological Explanations*. Berkeley: University of California Press.

Xenophon: Constitution of the Lacedaimonians. *La République des Lacédémoniens*, text and translation by François Ollier. Paris: F. Alcan, 1934.
Xenophon: *Memorabilia*. (*ed.*) J. Renick Smith. New York: Arno Press [1979].
Zahar, E. [1980]: 'Einstein, Meyerson and the Role of Mathematics in Physical Discovery', *British Journal for the Philosophy of Science* 31: 1 – 43.

NOTES ON CONTRIBUTORS

Robert Ackermann is a member of the Philosophy Department at the University of Massachusetts, Amherst. His book *The Philosophy of Karl Popper* was published in 1976.

Alan Chalmers is a member of the Department of General Philosophy at the University of Sydney. His book *What is This Thing Called Science?* first appeared in 1976.

Frank Cioffi is a member of the Department of Philosophy at the University of Essex. With Robert Borger he edited *Explanation in the Behavioural Sciences* which appeared in 1970.

L. Jonathan Cohen is a Fellow of The Queen's College, Oxford. His books on the philosophy of science were published in 1970 (*The Implications of Induction*) and 1977 (*The Probable and the Provable*).

Noretta Koertge is a member of the Department of the History and Philosophy of Science at Indiana University, Bloomington.

Richard Kraut is a member of the Department of Philosophy at the University of Illinois at Chicago. His book *Socrates and the State* is shortly to be published.

David Papineau is a member of the Department of the History and Philosophy of Science, Cambridge University. His book *For Science in the Social Sciences* was published in 1978.

Roland Puccetti is a member of the Department of Philosophy at Dalhousie University. His book *Persons* was published in 1968.

Alan Ryan is a Fellow of New College, Oxford. His book *The Philosophy of the Social Sciences* was published in 1970.

Wallis Suchting is a member of the Department of General Philosophy at the University of Sydney. His book *Marx. An Introduction* was published in 1983.

Peter Urbach is a member of the Department of Logic and Scientific Method at the London School of Economics.

Jeremy Waldron is a member of the Department of Politics at Edinburgh University.

INDEX OF NAMES

NIJHOFF INTERNATIONAL PHILOSOPHY SERIES

1. Rotenstreich N: Philosophy, History and Politics – Studies in Contemporary English Philosophy of History. 1976. ISBN 90-247-1743-4.
2. Srzednicki JTJ: Elements of Social and Political Philosophy. 1976. ISBN 90-247-1744-2.
3. Tatarkiewicz W: Analysis of Happiness. 1976. ISBN 90-247-1807-4.
4. Twardowski K: On the Content and Object of Presentations – A Psychological Investigation. Translated and with an Introduction by R. Grossman. 1977. ISBN 90-247-1726-7.
5. Tatarkiewicz W: A History of Six Ideas – An Essay in Aesthetics. 1980. ISBN 90-247-2233-0.
6. Noonan HW: Objects and Identity – An Examination of the Relative Identity Thesis and Its Consequences. 1980. ISBN 90-247-2292-6.
7. Crocker L: Positive Liberty – An Essay in Normative Political Philosophy. 1980. ISBN 90-247-2291-8.
8. Brentano F: The Theory of Categories. Translated by R.M. Chisholm and N. Guterman. 1981. ISBN 90-247-2302-7.
9. Marciszewski W (ed): Dictionary of Logic as Applied in the Study of Language – Concepts / Methods / Theories. 1981. ISBN 90-247-2123-7.
10. Ruzsa I: Modal Logic with Descriptions. 1981. ISBN 90-247-2473-2.
11. Hoffman P: The Anatomy of Idealism – Passivity and Activity in Kant, Hegel and Marx. 1982. ISBN 90-247-2708-1.
12. Gram, MS: Direct Realism – A Study of Perception. 1983. ISBN 90-247-2870-3.
13. Srzednicki, JTJ and Rickey, VF (eds): Leśniewski's Systems – Ontology and Mereology. ISBN 90-247-2879-7.
14. Smith, Joseph Wayne: Reductionism and Cultural Being – A Philosophical Critique of Sociobiological Reductionism and Physicalist Scientific Unificationism. 1984. ISBN 90-247-2884-3.
15. Zumbach C: The Transcendent Science – Kant's Conception of Biological Methodology. 1984. ISBN 90-247-2904-1.
16. Notturno, MA: Objectivity, Rationality and the Third Realm: Justification and the Grounds of Psychologism – A Study of Frege and Popper. 1984. ISBN 90-247-2956-4.
17. Dilman, I: Philosophy and Life. 1984. ISBN 90-247-2996-3.
18. Russell, Joseph J: Analysis and Dialectic. Studies in the Logic of Foundation Problems. 1984. ISBN 90-247-2990-4.
9. Currie, G and Musgrave, A: Popper and the Human Sciences. 1985. ISBN 90-247-2998-X.